D0069416

WILDFLOWERS
AND WEEDS

WILDFLOWERS AND WEEDS

A Field Guide in Full Color

BOOTH COURTENAY & JAMES H. ZIMMERMAN

PRENTICE
HALL
PRESS

New York London Toronto Sydney Tokyo Singapore

PRENTICE HALL PRESS
15 Columbus Circle
New York, NY 10023

PRENTICE HALL PRESS and colophons are registered trademarks of
Simon & Schuster, Inc.

Library of Congress Cataloging-in-Publication Data

Courtenay, Booth.
    Wildflowers and weeds : a field guide in full color / Booth Courtenay,
James H. Zimmerman.
        p.   cm.
    Reprint. Originally published: New York : Van Nostrand Reinhold,
1972.
    ISBN 0-13-957630-4 : $12.95
    1. Wildflowers—Great Lakes Region—Identification.   2. Weeds—
Great Lakes Region—Identification.   3. Wildflowers—Great Lakes
Region—Pictorial works.   4. Weeds—Great Lakes Region—Pictorial
works.   I. Zimmerman, James Hall, 1924-   . II. Title.
[QK130.C68   1990]
582.13'0977—dc20   90-7052
                                                                    CIP

Designed by Booth Courtenay

Manufactured in Japan

10 9 8 7 6 5 4 3 2 1

First Prentice Hall Press Edition

CONTENTS

WILDFLOWERS AND WEEDS

A man must see before he can say.
—Thoreau

This workbook is designed to help you *see* the wildflowers, those surviving without, or in spite of, man's hand.

Many may simply employ its wide range of color photographs to match picture with collected bloom. But this random, and often time-consuming, method of identification fails to help you understand how important plant character and site are in making sure of what you have found.

A more orderly approach offers faster recognition of individuals, greater knowledge of plant life in general. For those who would take up the challenge of learning to see, we give you two tools: HABITAT, or place of growth; and STRUCTURE, or physical characteristics. Use them; let the guide work for and with you. The earth's colorful floor will become increasingly a part of you, and you of it.

HABITAT

The large plant provinces are broken up into many divisions: from wet to dry, from sun to shade, from north to south. Only plants sharing similar needs can occur together within the divisions, or communities. Awareness of natural bedfellows leads to assurance of what to expect in varying environments, wherever you may be. (The Purple Cinquefoil, long adapted to dampness, will never be found in a dry open field near Pearly Everlasting, a typical dry-meadow plant.) The drawing on page **xii** and word sketches describe the ancient, and current, major communal homes, with a few of their typical residents.

STRUCTURE

Dawning curiosity about wildflowers is frequently stifled by the vast bewildering variety, even in one meadow or woods. It is an impossible task to memorize by rote hundreds of apparently unrelated individual flowers. Instead, in the second section, we have grouped plant families, with shared structural features (as number of petals or type of blossom), into an easily-remembered framework of eleven charts.

This FAMILY-GROUP system points up structural differences, as well as similarities, using only conspicuous family field traits. It requires neither microscope nor technical background.

In the pictorial glossary, drawings, not words, define the types and arrangements of leaves and flowers, and those parts to which we refer. For easier understanding, we have deliberately condensed, translated (or even omitted) scientific terms.

PLANTS INCLUDED

Vegetation is roughly divided into three layers: the canopy, or woody trees; the middle story, or woody shrubs and vines; and, at ground level, the non-woody plants, or HERBS, those that die back to the ground each year regardless of their life span—annual, biennial, or perennial.

The subjects here are the herbs, excluding grass-like ones and most waterweeds. There are, however, some of the low-growing, seeming almost herb-like, shrubs. The selection is a thoughtfully balanced cross section of the plants in the region. Choices were made by checklist studies of New York, Ohio, Indiana, Illinois, Iowa, Minnesota, Wisconsin, Michigan, and those covering southern Ontario and southwestern Quebec. (At the outermost borders of our range the coverage will, perforce, be less complete.)

The common names are those with, hopefully, the greatest acceptance. Unlike birds, plants have not had their vernacular names successfully standardized. To list the myriad variations of local and international origin takes a book in itself. (Just a few other names for what we call Wild Geranium are Wild Cranesbill, Cranesbill, Spotted Cranesbill, and Grandmother's Bonnet.)

Authority for the Latin, the internationally accepted and understood double name, is "Manual of Vascular Plants of the Northeastern United States and Adjacent Canada," Gleason and Cronquist (Van Nostrand Reinhold, 1963).

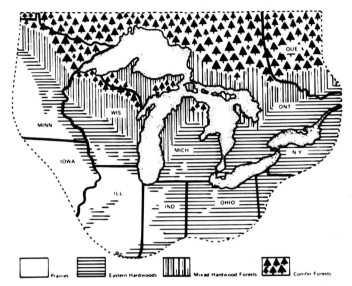

GEOGRAPHICAL RANGE MAP

REGION COVERED

There is a rich and diverse flora within our geographic range. It includes parts of four great plant provinces: Eastern Hardwoods, Mixed Hardwood Forests, Conifer Forests, and the open Prairies, or Grasslands. Here, where the four flow together, flourish a tremendous variety of native plants joined by an array of foreigners from all over the world.

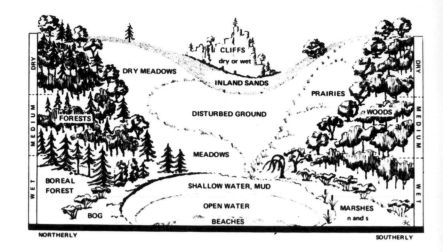

TOOL 1: HABITAT

This simplified representation of our major plant communities shows their progression from wet to dry, from dense shade to full sun, from north to south.

Neither state nor national boundaries mark them. Occasionally the breaks between them are sharp and easily seen: cliffs, hills, old burns, waterways, broad highways, or extensive cultivation. More commonly, there is only a subtle gradation, as one or more environmental factors (wetness, dryness, available sunlight, temperature, or soil character) have changed or are changing. Even a single stretch of forest may include wet, medium, and dry sections—with accompanying variations in plant forms or abundance.

In all of them are found assemblages of plants which may differ in ancestry, but which tend to occur together. They require, or tolerate, similar sites. Frequently they need the presence of some of the others to thrive. Prairies are not JUST grasses, nor forests ONLY trees. They are complexes of hundreds of competing, dependent, or cooperating organisms—herbs, grasses, trees, shrubs, vines, fungi, creatures—each perfectly adapted for the survival of its own species.

This adaptation, or evolution, has taken place in relatively short-term geologic time periods, in which a single kind of plant can change to accommodate to altering conditions.

The Cinquefoils, for example, have diversified in many ways to be able to exist in a wide variety of places: beaches, sunny prairies, disturbed ground, wet meadows, bogs, northern dry forests, and southern woods. Today, isolated in these differing communities, each individual has its *own* developed plant form, the one best suited for continued existence in its current habitat. (Some, less limited in tolerance, are not such specialists as to be confined to only one community— Wild Strawberry, for one.)

Association of flowers with their communities answers not only 'where' but 'why'. For greater ease of identification, as well as re-recognition, relate your finds to their homes: wetlands, woodlands, grasslands, sunny-dry or -damp; shady-dry or -damp; sand, rock, or heavy soils. The familiars, or their close cousins, will appear together in similar habitats wherever you travel.

WETLANDS

OPEN WATER   A still pool, a back eddy ignored by the flowing river, the sunny bays of inland lakes—these are the quiet, undisturbed waters that harbor the Water-lilies and other handsome, if less dramatic flowers.

SHALLOW WATER AND MUD   In sunny sloughs, on streambanks, and lake edges live plants whose roots demand a steady supply of water. Often found are the Water Smartweeds, Water Hemlock, and Marsh Milkweed.

MARSHES   The words 'marsh' and 'swamp', frequently used interchangeably, are not exactly synonymous. The swamp is an overflown area usually heavily wooded. The marsh, equally wet, is open and sunny, without tree growth. (As it often dries out as the season progresses, mud-loving plants, as Bur Marigold, may appear.) Cat-tail and the fragrant Sweet Flag are typical neighbors.

BOG The sunny open bog of our north country (with a few relics southward) is a spongy wet layer, varying in thickness, floating on an ages-old pocket of water. Eventually the soft carpet of the masses of dead Sphagnum Moss and other bog plants becomes solid enough to support heavier growth. The scattered small Tamaracks and Birches seen above the ground layer are forecasters of the Wet Forest to come. With the many Blueberry Family members grow some of our rare orchids, and the insect-eating Pitcher Plant and Sundew.

WOODLANDS

WOODS In earlier days, leafy hardwood forests stretched from the Atlantic to the Great Plains. For today's remnants, we have chosen to call them 'woods', for that does not connote the wilderness now gone. Tree dominants mark each of the three principal divisions.

> WET With Silver Maple, Willows, and Elms are commonly found Wood Nettle, Blue Marsh Violet, and Skunk Cabbage.
>
> MEDIUM Among the Sugar Maples and Basswood grow Bloodroot and False Rue Anemone.
>
> DRY Here with Oaks, Hickory, and Cherry are Solomon's Plume, Sweet Cicely, and Wild Geranium.

FORESTS The more northern heavily wooded communities, all of which include evergreens, can be separated into four major groups:

> BOREAL Balsam Fir, Arbor Vitae, White Pine, White Spruce.
>
> WET Tamarack, Black Spruce.
>
> MEDIUM Hemlock, Yellow Birch, Sugar Maple.
>
> DRY Jack Pine, Red Pine, Poplar.

The soil—wet to dry, thin sand or gravel to deep clay or loam—and the occurrence of fire determine the dominants of each. While their over-all appearance differs from one to another, many of the same flowers will be found in all of them, with a variable only of abundance.

Among the typical of the northern plants are Bunchberry, Yellow Beadlily, and Largeleaf Aster.

GRASSLANDS

PRAIRIES   The earliest plant-hunters, leaving the eastern hardwoods on their treks to the west, found a magnificent waving sea of plant life completely new to European eyes: the Prairie. Today it has all but vanished into plowed fields and villages and towns, as man has made use of its rich soil. In the absence of the burnings necessary for their survival, Midwestern upland prairies succumbed to woods invasion—but there are still remnants of these native communities. They require full sun, abhor soil disturbance or intensive grazing, and vary from wet to dry.

>WET   With the characteristic Tall Cord grass, frequent flowers are the New England Aster, Culver's Root, tall Meadowrue.

>MEDIUM   Here Bluestem grasses share living space with the Showy Sunflower, Blazing Stars, Smooth Aster, and the now so rare Compass Plant.

>DRY   Scattered among the short Grama grasses are the seasonal changes of the early spring Pasque Flower, midsummer Lead Plant with its soft gray foliage, and the fall Silky Aster.

WET MEADOWS   The meadows, a step up from the water, and sometimes called 'dry marshes', are covered with sun-loving plants which need damp soil. With no fire or mowing, they are invaded by thicket-forming shrubs. With the rustling sedges and grasses of this treeless community grow Marsh Marigold, Wild Iris, and fragrant mints.

DRY MEADOWS   Thick sod-forming grasses take over such formerly disturbed places as abandoned pastures and croplands, old forest clearings or burnings. In the resulting well-drained, sunny fields are Wild Strawberry, Pearly Everlasting, and Black-eyed Susan.

RAPID-CHANGE SITES

DISTURBED GROUND To the botanist a weed is not just something growing where he doesn't want it, though it may well be that, too. He describes it as a "rapidly-growing herb, frequently an annual, that can start to grow only in freshly disturbed ground, usually sunny." This disturbance, or turning-over of soil, can be by wind, erosion, water-lowering; by plow, bulldozer, shovel and hoe, or grazing. Thus defined, this large group of native and foreign plants includes our cereal grains, corn, and most crop and garden plants. Some of the 'wild' ones, of less economic value, are Dandelion, White Campion, Ragweed, Yarrow.

BEACHES The sandy shores of our lakes, living water-ways, grow or diminish or alter as climate and water levels vary. The basic struggle here is not one plant with another, but with the ceaseless shifting of the sands and the drying of the sun. Beach grasses, Silverweed, and Beach Pea are examples of successful adaptation to a difficult home-place.

INLAND SANDS 'Barrens' gradually develop on relic dunes of ancient lake and river beds. These stabilized-sand regions support the widely-spaced growth of such trees as Jack Pine and Scrub Oak. In the sunniest parts are plants from neighboring open lands—Wormwood, Flowering Spurge, Wild Rose. As shade increases flowers of the Dry Forest or Dry Woods take over.

CLIFFS The 'rapid-change' here is not time, but space— inch to inch: a *fraction* more or less of ground in which to spread roots, a *bit* more or less of moisture. Those plants particularly suited to finding nourishment in the thin layer of soil over the rocks are able to live in this rugged habitat—although many of them may be found in other communities as well. With Birch, Pine, Hemlock, and a ground cover of mosses, lichens, and ferns grow Columbine, Rock Cress, Harebell.

TOOL 2: STRUCTURE

We have noted in the Habitat section that a *single* kind of plant may become a *group* of plants in relatively short-term geologic periods—a few million years.

But over a much longer time-span there have been major changes in land masses, oceans, mountains, temperature, and climate. These have either wiped out parts of the plant world, or allowed it to experiment its way into radically new forms. The surviving lines of descent, some primitive, some much altered, today constitute the evolutionary tree-of-life. This 'tree' of structural kinship provides logical basis for orderly arrangement of all life forms.

To the scientist the most useful indicators of relationship are the fine details of the reproductive parts and systems, and the internal anatomy. These primary features are less quickly modified by habitat differences than are the secondary, or external, characters such as leaf shape, flower color, or growth habit. Being more stable, they provide a firm base for classification into families— the largest limbs of the evolutionary tree.

FAMILY

This is a convenient major grouping of all those plants that share a *combination* of enough distinctive features to suggest common ancestry. Some families are closely knit, with strong resemblances among most of the relatives.

All members of the Mustard family have four petals, six stamens, are herbaceous. Most plants in the Parsley family bear the clear stamp of kinship: umbelled flowers, sheathing leaves.

Some seem widely different.

The Rose family ranges from the low-growing Strawberry, through the Rasberries, up to the woody Apple tree. But though they may vary greatly in one or more characters, they share certain traits: leaf appendages and joined sepals.

Some appear smaller and more uniform than they actually are, when but few representatives live in our part of the world.

The Iris family has a much greater variety of forms in Africa than in North America.

Other families are in fact small: a few plants so different from all other groups that they rate 'family' status. In some instances these may be relic survivors of a once larger association.

An extreme of this is the Lopseed family, erected to accommodate a single plant, looking like a Mint, but having very un-mint-like fruit.

GENUS

The genus, a first branching of the large limb, is a more geologically recent division of the family. It is a GROUP of KINDS of plants that share very many basic indicators of relationship, but differ in external details—leaf shape, flower color or size, habitat.

A look at the Cinquefoils (pp. 39, 40) shows the strong flower similarities as well as the differences in growth habit and where they may be found.

SPECIES

This is the twig of the tree-of-life, a subdivision of the genus. It is a relatively uniform INDIVIDUAL plant KIND, distinct enough in structure and behavior to be recognized as a named entity.

Silverweed (p. 39) is the only Cinquefoil with the special character-combination of long, above-ground runners and silvery feather-divided leaves that is found on beaches and inland sands.

Its Latin name, *Potentilla anserina,* shows: 1. GROUP kinship—genus *Potentilla;* 2. its uniqueness—species *anserina.* It is by this name that it can be known anywhere in the world, regardless of local language. (This one happens to be a common plant in Europe and England, as well as in North America—and will be found in similar habitats there as well as here.)

HOW TO USE THE FAMILY-GROUP CHARTS

Fortunately for the amateur, there are usually enough external features to identify a plant. Using these easily noted characters of petals, sepals, leaves, the plant families are divided into eleven GROUPS—charts A through K. (For a definition of terms, see the two-page pictorial glossary.)

Each chart, headed by number of petals or type of tubular flower, begins with *one* Type Family with many members in it. This is followed by *several* Associated Families.

In Chart A, the Lily family, with its 3 PETALS, is the 'Type'. All of the accompanying Associates *also* have three petals.

The important external features of the Lily family are described. For the Associates, however, *only* those characters which *differ* from the 'Type' are mentioned.

The Lily's petals are all alike, whereas *one* of the Orchid's three is very different from the other two. This, then, is noted. The leaves, being basically similar, are not.

To start the identity search, count the number of petals. Notice whether they are separate or joined together in a tube. (Look at several flowers to be sure you have one that developed normally, that has not been attacked by insects, or one whose petals have not started to fall.)

Then check the eleven 'Type' families. Often this is all you need to find the appropriate chart.

There *are* times when you need further clues: number of sepals, stamens, the kind or arrangement of leaves. If the flower is not too tiny, note whether the seed capsule is inside the blossom or just beneath, appearing as a swelling in the flower stalk. Of further importance may be the character of the plant itself: matting, creeping, or climbing; aromatic when crushed or broken; in patches or clumps.

PICTORIAL GLOSSARY

FLOWER PARTS

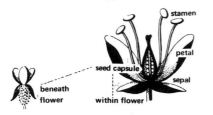

stamen

petal

seed capsule

sepal

beneath flower

within flower

BUD, the future shoot for either leaves or flowers

INFLORESCENCE – FLOWER ARRANGEMENT

umbel

solitary

head

spike

raceme

cyme

LEAF TYPES

parallel-veined

untoothed

toothed

lobed

palmately lobed

feather lobed

sheathing

basal appendage
may be variously shaped

divided

palmately divided

feather divided

LEAF ARRANGEMENT

basal rosette

whorled

alternate

opposite

FLOWER TYPES, keyed to Family Grouping Charts

3 PETALS, 3 SEPALS Chart A

petals and sepals ALIKE

petals DIFFERENT from sepals

1 of 3 petals VERY different

4 PETALS Chart B

seed capsule INSIDE flower

seed capsule BENEATH flower

5 PETALS

★ WITH OPPOSITE LEAVES Chart C

sepals occ. joined to form a tube

PETALS DIFFERENT Chart F

bilateral symmetry

spur

★ SEPALS SEPARATE Chart D

sepals often petal like

★ FLOWERS TINY — IN UMBELS Chart G

★ SEPALS JOINED Chart E

deeply or shallowly

★ FLOWERS TINY — VARIOUSLY ARRANGED
Chart H

★ these have radial symmetry

FLOWERS TUBULAR

RADIAL SYMMETRY Chart I

BILATERAL SYMMETRY Chart J

BOTH RADIAL and BILATERAL SYMMETRY Chart K

radial

bilateral

arrangement in head

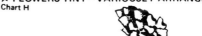

With the additional observations, you can do a closer reading of the selected chart. Examine the 'Type' first—and then the Associates—for the family with the combination most nearly matching your flower specimen.

With *family* choice made, turn to it in the color section for *genus* and *species* photographs and descriptions.

If, at times, you miss, look again at the character possibilities. It is rarely by *one* feature alone that a plant can be accurately identified. (If you have found a flower we were unable to include in our species selection, be assured we have pictured a very close relative.)

REMEMBER ALWAYS: it is the COMBINATION of CHARACTERS that makes identification CERTAIN.

EXAMPLES

1. Beside the road is a bright blue flower with THREE separate PETALS, behind which are THREE separate green SEPALS. Tiny hairs on the yellow stamens give the center a fuzzy look. The slender PARALLEL-VEINED LEAVES seem to sheathe the stem and the lower part of the flower cluster, which is an UMBEL.

In only one chart, A, are there '3 PETALS, 3 SEPALS',— the Lily Group. Although the plant almost fits the word description of the 'Type', you may suspect that this bright blue blossom is not, in fact, a lily. Look at the Associates. The 'fuzzy stamens and sheathing leaves' make the Spiderwort family a promising candidate. Page 2 of the photographic section will affirm your judgement with a picture of one of our common wayside Spiderworts, *Tradescantia ohiensis*.

2. In woodland shade you come upon a plant with a LOOSE CLUSTER of TUBULAR flowers, each RADIALLY SYMMETRICAL and 5-lobed. The STAMENS extend somewhat beyond the tube, giving the blossom a soft look. The LEAVES are ALTERNATE on the stem and deeply LOBED. The plant is a bit HAIRY.

There are two charts with radially tubular flowers, I and K. But K has flowers crowded into TIGHT HEADS, and yours are in a loose cluster. Therefore you choose

I, the Bluebell Group. The 'Type' family does not have lobed leaves; so you check the Associates, remembering that the leaves are *alternate* as well as lobed, and the plant *hairy*. The Waterleaf family seems to have all the characteristics you have noted. The photograph, p. 89, will show you which species you have found.

3. At the edge of a cornfield is a one-inch wide, cream-colored, mahogany-centered flower with 5 PETALS, all ALIKE and SEPARATE. It has a thick yellow COLUMN in the middle; the back of the flower seems enclosed in a transparent cup—these are JOINED SEPALS. The LEAVES are ALTERNATE and PALMATELY LOBED.

This detecting requires a little more care, since there are four charts with '5 PETALS, ALIKE, SEPARATE' (C, D, E, G). But Chart C has OPPOSITE leaf arrangement; and G has very small flowers.

A close look at the plant shows a pair of small AP-PENDAGES at the base of each leaf-stem. In only chart E, the Rose Group, are 'leaf appendages' and 'joined sepals' 'Type' family characters in combination. Here, among the Associates, the Mallow family is the one with the peculiar central column. A look at p. 44 will prove your flower to be a 'disturbed ground' mallow, Flower of an Hour, *Hibiscus trionum.*

EXPERIENCE

Only continued and repeated OBSERVATION of plant as well as blossom can make the wildflowers familiar. The charts provide important guidelines—field characteristics for which to look. But it must be your own eyes that do the watchful work. The more practice you have, the easier the process becomes.

Later, with growing experience, perception, and desire for more detailed knowledge, you will find it rewarding to go from this book-of-the-field to the technical and definitive textbooks.

GOOD HUNTING.

CHART KEY

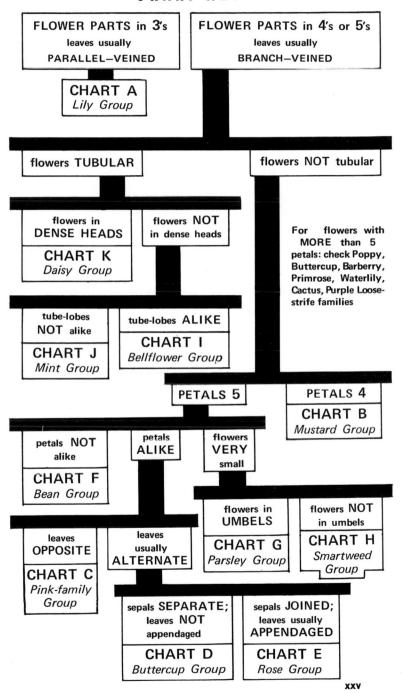

FLOWER PARTS in 3's
leaves usually
PARALLEL—VEINED

CHART A
Lily Group

FLOWER PARTS in 4's or 5's
leaves usually
BRANCH—VEINED

flowers **TUBULAR**

flowers **NOT** tubular

flowers in
DENSE HEADS
CHART K
Daisy Group

flowers **NOT**
in dense heads

For flowers with **MORE** than 5 petals: check Poppy, Buttercup, Barberry, Primrose, Waterlily, Cactus, Purple Loosestrife families

tube-lobes
NOT alike
CHART J
Mint Group

tube-lobes **ALIKE**
CHART I
Bellflower Group

PETALS 5

PETALS 4
CHART B
Mustard Group

petals **NOT**
alike
CHART F
Bean Group

petals
ALIKE

flowers
VERY
small

flowers in
UMBELS
CHART G
Parsley Group

flowers **NOT**
in umbels
CHART H
Smartweed Group

leaves
OPPOSITE
CHART C
Pink-family Group

leaves
usually
ALTERNATE

sepals **SEPARATE**;
leaves **NOT**
appendaged
CHART D
Buttercup Group

sepals **JOINED**;
leaves usually
APPENDAGED
CHART E
Rose Group

3 PETALS, 3 SEPALS

Type family:

LILY, *Liliaceae* 4—9

PETALS — 3, alike
SEPALS — 3, alike; color of petals (exc. Trillium)
STAMENS — 6
FLOWERS — in a raceme, spike, umbel, or solitary; occ. a bell or funnel of petals and sepals united into a tube
SEED CAPSULE — 1, inside flower, becoming a berry or dry case
LEAVES — parallel-veined; broad to grass-like, untoothed, undivided; alternate or whorled, often sheathing
PLANT — erect (climbing in some *Smilax*)

ASSOCIATED FAMILIES

INDIVIDUAL FLOWERS DISTINCT

WATER—PLANTAIN, *Alismataceae* 1 SEPALS — green STAMENS — 6 or more SEED CAPSULES — many, in ball-like mass LEAVES — narrow to arrow-shaped	**PICKEREL WEED,** *Pontederiaceae* 2 PETALS — somewhat joined FLOWERS — funnel-shaped, in a crowded spike just above sheathing leaf LEAVES — heart-shaped
SPIDERWORT, *Commelinaceae* 2 PETALS — 1 of 3 may be small SEPALS — green, occ. hairy STAMENS — may be fuzzy FLOWERS — in umbels, clasped by sheathing leaves	**FLOWERING RUSH,** *Butomaceae* 3 STAMENS — 9 FLOWERS — in umbels SEED CAPSULES — 6 LEAVES — slender
YELLOW—EYED GRASS, *Xyridaceae* * FLOWERS — crowded in short brown cone-like structures at stem tip LEAVES — slender	**ARROW GRASS,** *Juncaginaceae* * FLOWERS — small, greenish, in spikes SEED CAPSULES — 3 or 6 LEAVES — slender

SEED CAPSULE BENEATH FLOWER

AMARYLLIS, *Amaryllidaceae* 9 FLOWERS and stalks hairy LEAVES — grass-like	**ORCHID,** *Orchidaceae* 11—16 PETALS — 1 of 3 VERY different SEPALS — 2 or 3 may be joined STAMENS — 1 or 2
IRIS, *Iridaceae* 10 STAMENS — 3, hidden LEAVES — folded, forming flat sheathing fans	**YAM,** *Dioscoreaceae* 16 FLOWERS — small, in racemes PLANT — twining wine

*not illustrated

PETALS and SEPALS ALIKE,
sepals colored like petals

PETALS and SEPALS DIFFERENT,
sepals green

INDIVIDUAL FLOWERS TINY, CROWDED	
CAT—TAIL, *Typhaceae* 1 FLOWERS — minute, packed on long thick spike LEAVES — slender, D—shaped in cross-section	**PIPEWORT,** *Eriocaulaceae* 2 FLOWERS — in tiny white buttons topping leafless stems LEAVES — in basal rosette
BUR—REED, *Sparganiaceae* 1 FLOWERS — in ball-like clusters on zig-zag stem LEAVES — slender, triangular in cross-section	**CALLA,** *Araceae* 3 FLOWERS — packed on a thick spike above or inside a sheathing green or colored leaf LEAVES — not always parallel-veined

NOTES

Some members of the following families may have 3- or 6-parted *conspicuous* flowers, but NOT parallel-veined leaves: Buttercup, Waterlily, Barberry, Milkwort, Purple Loosestrife, Gourd, and Birthwort. Some with *inconspicuous* flowers may be found in the Rockrose, Smartweed, False Mermaid, Nettle, Crowberry, and Bedstraw families.

Families in which parallel-veined leaves MAY occur (but flowers are not 3-parted) are: Pink, Saxifrage, Parsley, Plantain, Daisy.

Occasionally a 3-petaled flower may appear as a 'sport' on a normally 4- or 5-parted species. If possible, observe several plants to be certain you haven't found an 'odd' one.

4 PETALS

Type family:

MUSTARD, *Cruciferae* 17—20

PETALS — 4, separate, alike, or nearly so
SEPALS — 4, separate, alike
STAMENS — 6 (2 short)
FLOWERS — in elongating terminal raceme or spike
SEED CAPSULE — 1, inside flower, from slender to round, with persistent central partition
LEAVES — alternate; undivided or feather-divided or lobed; some in basal rosette

ASSOCIATED FAMILIES

LEAVES LOBED or *DIVIDED*

CAPER, *Capparidaceae* 16
STAMENS — 8 or more
LEAVES — palmately divided

POPPY, *Papaveraceae* 20
PETALS — 4, 8, or 12
STAMENS — more than 10
FLOWER — may be solitary
PLANT — has white, yellow, or red juice

BARBERRY, *Berberidaceae* 20
PETALS — 6: leaves divided (Blue Cohosh)
6—9: divided leaf forms umbrella over flower (Mayapple)
8: leaves 2—divided (Twinleaf)
STAMENS — more than 10
FLOWERS — solitary in Mayapple, Twinleaf

BLEEDING HEART, *Fumariaceae* 21—22
PETALS — not all alike, joined at tip; 1 or 2 with nectar sac or spur
LEAVES — finely divided, blue-green

LEAVES UNDIVIDED and *USUALLY UNLOBED*

EVENING PRIMROSE (in part), *Onagraceae* 74—75
STAMENS — 2—8
SEED CAPSULE — beneath flower
LEAVES — opposite OR alternate

MEADOW BEAUTY, *Melastomaceae* 21
SEPALS — form cup enclosing seed capsule
STAMENS — 8
LEAVES — opposite

NOTES

Other families with members which may seem to have 4 (or 8) separate petals;
1. leaves NOT lobed or divided, often *opposite:* Blueberry, Bedstraw, Dogwood, Pink, Purple Loosestrife, Nettle, Mulberry, Saxifrage, Smartweed, Mistletoe, Snapdragon, Sedum;
2. leaves lobed or divided, usually *alternate:* Buttercup, Waterlily, Rose;
3. leaves *parallel-veined:* Plantain, Lily (Wild Lily of the Valley). Occasionally a 3 or 5 petaled flower may occur as a 'sport' with only 4.

CHART C *The Pink-Family Group*

5 PETALS—ALIKE, SEPARATE

(leaves opposite or whorled)

Type family:

PINK, *Caryophyllaceae* 22—25

PETALS — 5, alike, separate: notched (Chickweed type)
or bent out to form a flat face(Campion type)

SEPALS — 5, alike, separate (exc. in Campion type,
which are joined into a tube or sac)

STAMENS — 5 to 10

FLOWERS — in cymes

SEED CAPSULE — 1, inside flower

LEAVES — opposite or whorled; usually untoothed, not lobed or divided; occ.
parallel-veined; occ. appendaged

STEM — often leafy; often slightly swollen where leaves attach

 Chickweed

 Campion

ASSOCIATED FAMILIES

ST. JOHN'S—WORT,
Hypericaceae 26—27

STAMENS — many, conspicuous

LEAVES — often with minute black
or translucent dots

PURPLE LOOSESTRIFE,
Lythraceae 26—27

PETALS — 3—7, attached to sepal
tube

SEPALS — form tube enclosing seed
capsule

FLOWERS — in a spike, or grouped
along stem

LEAVES — occ. alternate on upper
stem

MILKWEED,
Asclepiadaceae 27—28

PETALS and SEPALS — turned
down (lower story)

STAMENS and NECTAR HOODS —
turned up (upper story)

FLOWERS - in umbels

SEED CAPSULE — hidden in upper
story, becoming a pod with silky
parachuting seeds

PLANT - with milky juice (exc.
Butterfly Weed)

PRIMROSE (in part),
Primulaceae 84—85

PETALS - open wide; occ. slightly
joined at base; 7 in Star Flower

FLOWERS — variously arranged

NOTES

Other families with members which seem to have 5 separate petals or sepals AND
opposite or whorled leaves:

1. stems with ONLY 1 or a FEW SETS of opposite or whorled leaves, other
leaves at base: Buttercup, Saxifrage, Geranium, Purslane;

2. stems LEAFY, with MANY SETS of opposite or whorled leaves; flowers
small: Milkwort, Sedum, Amaranth, Nettle, Goosefoot, Carpetweed, Mulberry,
Daisy (Quickweed).

5 PETALS—ALIKE, SEPARATE
(sepals separate)

Type family:

BUTTERCUP, *Ranunculaceae*   30—35

PETALS and/or SEPALS — 5 (occ. more), alike, separate; fall before seeds ripen

SEPALS — petal-like if petals are lacking

STAMENS — many, usually over 10

SEED CAPSULES — small, several to many, inside flower

LEAVES — alternate; usually divided, lobed, or toothed; bases often slightly sheathing stem

ASSOCIATED FAMILIES

seed capsule 1 (exc. Sedum); sepals may persist; leaves undivided, untoothed

WATERLILY, *Nymphaeaceae*   35
PETALS — often more than 5
LEAVES — rounded

PLANTS SUCCULENT

PURSLANE, *Portulacaceae*   28
PETALS — soon shriveling
SEPALS — 2
STAMENS — 5 to 20
SEED CAPSULE — may be beneath flower

CACTUS, *Cactaceae*   36
PETALS — more than 5
SEED CAPSULE — beneath flower
STEMS — often spiny; occ. tiny leaves

SEDUM, *Crassulaceae*   29
PETALS — 5 or 4 (Ditch Stonecrop)
SEPAL — 5 or 4
STAMENS — 8 to 10
SEED CAPSULES — 5 or 4
LEAVES — alternate, opposite, or whorled

PETALS FALL EACH DAY

ROCKROSE, *Cistaceae*   36
SEPALS — 2 of 5 small or lacking
FLOWER — opens only in full sun
STEM — wiry, leafy

FLAX, *Linaceae*   35
STAMENS — 5
LEAVES — slender
STEM — leafy

NOTES

Other families some of whose members may have more than 5 petals or petal-like sepals: Poppy, Barberry, Purple Loosestrife, Primrose, Lizard Tail.

The small (to 8″) musk-scented Moschatel, *Adoxa*, (not illustrated) has 3-divided leaves and a few 4- or 5-petaled green flowers.

CHART E   *The Rose Group*
5 PETALS—ALIKE, SEPARATE
(sepals joined)

Type family:
ROSE, *Rosaceae*   37-41

PETALS — 5, alike, separate

SEPALS — 5, joined slightly at base to form a persistent cup or disc (occ. 5 sepal-like leaves behind flower make it appear 10-sepaled)

STAMENS — many, over 10

SEED CAPSULE — 1 to many; inside flower (beneath in Wild Rose)

LEAVES — alternate; usually divided or lobed; sharply toothed; usually with 2 APPENDAGES at base of leaf stalk

ASSOCIATED FAMILIES

seed capsule 1; stamens only 5-10 (exc. Mallow, Pitcher Plant)

LEAVES LOBED or *DIVIDED*

WOOD—SORREL,
   Oxalidaceae 41

LEAVES — clover-like, but not toothed; occ. no appendages

SEED CAPSULE — upright pod

SAXIFRAGE (in part),
   Saxifragaceae 42—43

PETALS — may be shorter than sepals

SEED CAPSULE — 2-lobed

LEAVES — usually no appendages

GERANIUM,
   Geraniaceae 42

LEAVES — palmately lobed or divided; alternate or opposite

MALLOW, *Malvaceae* 44—45

STAMENS — joined to form conspicuous column

SEED CAPSULE — button-like, or oval pod

LEAVES — often palmately lobed, occ. rounded

LEAVES UNLOBED and *UNTOOTHED*

in basal rosette; no appendages (exc. Sundew)

SAXIFRAGE (in part),
   Saxifragaceae 42—43

LEAVES — with hairs, or rounded

SUNDEW, *Droseraceae* 45

FLOWERS — small, opening 1 or a few at a time

LEAVES — have sticky-tipped hairs

PITCHER PLANT,
   Sarraceniaceae 45

SEPALS — petal-like, persisting long after fall of petals

LEAVES — pitcher-like tubes

BLUEBERRY (in part),
   Ericaceae 75—79

PLANT — non-green (white to red, brown, black) or evergreen

PRIMROSE (in part)
   Primulaceae 84—85

PETALS — turn back

FLOWERS — droop; in umbels

LEAVES — tongue-shaped, hairless

NOTES

Petals joined but appearing separate occur in the Tomato family. *Sepals* slightly joined, and persistent, may be found in some families of Chart D. Hepatica (Buttercup family) has 3 persistent *sepal-like* leaves behind the flower.

Appendaged leaves may also occur in: Pink, Buttercup (False Rue Anemone), Bean, Senna, Violet, Ginseng, Smartweed, Nettle, and Honeysuckle families.

Type family:
BEAN, *Fabaceae* 46—53

PETALS — 5, separate, not alike
SEPALS — 5, united in a tube free from other flower parts
STAMENS — 10, usually united into a tube
FLOWERS — have bilateral symmetry; in racemes, heads, occ. spikes or umbels
SEED CAPSULE — inside flower; a flat or rounded pod
LEAVES — alternate; 3- or many-divided; NOT TOOTHED, exc. some Clovers;
2 APPENDAGES at base of leaf stalk

ASSOCIATED FAMILIES

VIOLET, *Violaceae* 54—56

PETALS — the lowest extends back
to form a nectar spur
STAMENS — 5, separate
FLOWERS — solitary
LEAVES — undivided; toothed;
often lobed

MILKWORT,
Polygalaceae 53—54

PETALS — 3, 1 different
SEPALS — 5, 2 enlarged, petal-like
STAMENS — 8
SEED CAPSULE - 2—seeded
LEAVES — undivided; no
appendages

SENNA,
Caesalpiniaceae 52—53

PETALS — almost alike
STAMENS — unequal, not united

JEWELWEED,
Balsaminaceae 57

FLOWERS — have large sac-like
spur; hang from slender stalks
SEED CAPSULE — explodes on
touch
LEAVES — undivided; toothed; no
appendages
STEM — smooth, watery-translucent

NOTE

Bilateral symmetry also occurs in the Buttercup (Delphinium, Monkshood),
Spiderwort, Pickerel Weed, Saxifrage (Alum Root), and Orchid families; and in
the Mint and Daisy groups.

5 PETALS—FLOWERS TINY *(in umbels)*

Type family:

PARSLEY, *Umbelliferae*   57–60

PETALS — 5, alike, separate
SEPALS — 5, small or none
STAMENS — 5
FLOWERS — in flat, domed, or balled umbels (occ. in heads)
SEED CAPSULE — beneath flower; dry; 2-seeded
LEAVES — alternate; usually divided (not Rattlesnake Master); often toothed;
STRONGLY SHEATHING around stem

ASSOCIATED FAMILIES

GINSENG, *Araliaceae* 61

STAMENS — 5 or 10
FLOWERS — in balled umbels
FRUIT — a berry
LEAVES — occ. with appendages

BUCKTHORN, *Rhamnaceae* 61

STAMENS — 5 or 10
FLOWERS — in oval clusters of umbels
SEED CAPSULE — 3-seeded; cup-like base persistent
LEAVES — undivided, not sheathing
PLANT — small woody shrub

SANDALWOOD, *Santalaceae* 61

PETALS — none
SEPALS — 5, petal-like
FLOWERS — star-like, in flat-top cluster
FRUIT — nut-like
LEAVES — untoothed, undivided, not sheathing

NOTES

Umbels may also occur in Flowering Rush, Spiderwort, Lily, Amaryllis, Primrose, Wood-Sorrel, Bean, Buttercup, and Milkweed families.

Sheathing leaf bases are also in some members of the Buttercup family and in the Lily group.

5 PETAL-LIKE SEPALS —FLOWERS TINY

(not in umbels)

Type family:
SMARTWEED, *Polygonaceae* 62—64

PETALS — none
SEPALS — 5 (rarely 3-6) green or colored, petal-like
FLOWERS — in short or long spikes, or racemes
SEED — 1
LEAVES — alternate; untoothed, undivided; with appendages uniquely joined
into closed papery sheath around stem, making it appear jointed.

ASSOCIATED FAMILIES

appendages only in Nettle family

LIZARD TAIL, *Saururaceae* 65
SEPALS — none
STAMENS — 6-7, white
FLOWERS — crowded, in curved
spike

NETTLE, *Urticaceae* 65
SEPALS — 2-5, green or whitish
FLOWERS — in clusters or spikes
LEAVES — alternate OR opposite;
toothed, undivided
STEM — watery or bristly

POKEWEED, *Phytolaccaceae* 66
SEPALS — 5, rounded; green-white
to purple
STAMENS — 10
FLOWERS — in raceme
FRUIT — 10-seeded berry

SPURGE,
Euphorbiaceae 66—67
SEPALS — none
FLOWERS — tiny, in groups tightly
bound by 2-6 petal-like leaves —
green white, or colored
FRUIT — 3-seeded
PLANT — often with milky juice

FALSE MERMAID,
Limnanthaceae 66
PETALS — 3, very tiny
SEPALS — 3, green
FLOWER — solitary
SEEDS — 1-3
LEAVES — divided into slender
segments

GOOSEFOOT,
Chenopodiaceae 67—68
FLOWERS — in small groups, or
balled clusters
LEAVES — lobed or toothed; occ.
opposite

AMARANTH,
Amaranthaceae 68
FLOWERS — in dense clusters, amid
tiny scaly leaves
LEAVES — lobed or toothed; occ.
opposite

LEAVES OPPOSITE or WHORLED

MULBERRY, *Moraceae* 69 LEAVES — palmately divided	**MISTLETOE,** *Loranthaceae* * FLOWERS — 1, at end of each short branch PLANT — tiny; reddish, parasite on twigs of some conifers
CARPETWEED, *Aizoaceae* 69 SEED CAPSULE — 3-celled LEAVES — whorled PLANT — mat-forming	Other families with mat-forming members: ALTERNATE LEAVES: Purslane, Sedum, Spurge, Smartweed, Amaranth; OPPOSITE LEAVES: Saxifrage, Chickweed, Evening Primrose, Primrose, Bedstraw, Verbena, Mint

WOODY VINES AND CREEPERS

Flowers hidden or in crowded clusters

SUMAC, *Anacardiaceae* 70 LEAVES — 3-divided FRUIT — hard berry, white or yellowish PLANT — erect, or climbing by aerial rootlets	**MOONSEED,** *Menispermaceae* * LEAVES — 3-7 lobed FRUIT — black berry, seed crescent-shaped PLANT — twining
GRAPE, *Vitaceae* * LEAVES — palmately lobed or divided FRUIT — berry PLANT — climbing by tendrils or rootlets	**BITTERSWEET,** *Celastraceae* * LEAVES — toothed; undivided, unlobed FRUIT — orange, 3-divided PLANT — twining
CROWBERRY, *Empetraceae* * LEAVES— like tiny spruce needles FRUIT — black berry PLANT — low, creeping, evergreen	Other small evergreen creepers are in Blueberry, Bedstraw, Honeysuckle families

NOTES

Small, inconspicuous, separate-petaled flowers may also be found in the Mustard, Plantain, Pink, Buttercup, Rose, Saxifrage, Bean, and Milkwort families

*not illustrated

CHART I *The Bellflower Group*

TUBULAR FLOWERS *(radially symmetrical)*

Type family:

BLUEBELL, *Campanulaceae* 71

PETAL LOBES — 5, alike
SEPAL LOBES — 5, alike
SEED CAPSULE — beneath flower; small; many-seeded
LEAVES — alternate; undivided, unlobed; may be toothed

ASSOCIATED FAMILIES

FOUR O'CLOCK, *Nyctaginaceae* 70

PETALS — none
SEPALS — form colored tube, behind which is a green cup of joined leaves
SEED — 1
LEAVES — opposite
STEM — swollen at joints

BEDSTRAW, *Rubiaceae* 71—72

PETAL LOBES — 3-4
FLOWER — often small
FRUIT — a tiny bur, or berry
LEAVES — opposite or whorled

HONEYSUCKLE, *Caprifoliaceae* 73

FLOWERS — often paired; occ. spurred
FRUIT — berry or small dry capsule
LEAVES — opposite; occ. appendaged
PLANT — often woody

GOURD, *Cucurbitaceae* 73

PETAL LOBES — 5-6
FRUIT — prickly, few-seeded
LEAVES — lobed
PLANT — non-woody vine with tendrils

EVENING PRIMROSE (in part), *Onagraceae* 74—75

PETAL and SEPAL LOBES — 2-4
SEED CAPSULE — 4-sided, often slender; or tiny bur
SEEDS — occ. have parachutes
LEAVES — opposite, alternate, or in rosette; occ. toothed or lobed

BLUEBERRY (in part), *Ericaceae* 75—79

PETAL LOBES — 4-5

SEED CAPSULE — beneath OR within flower
FRUIT — berry or small dry capsule
LEAVES — often evergreen; occ. opposite
PLANT — may be woody or trailing

DOGWOOD, *Cornaceae* 79

PETAL LOBES — 4, tiny
FLOWERS — in small cluster bound by 4 petal-like white leaves
FRUIT — berry
LEAVES — opposite or whorled

BIRTHWORT, *Aristolochiaceae* 79

FLOWER TUBE — 3-parted; on ground
LEAVES — heart-shaped, fuzzy

TEASEL, *Dipsacaceae* 79

PETAL LOBES — 4
FLOWERS — narrow tubes packed in domed or cylindrical heads
SEED — 1
LEAVES — opposite; lobed or feather-divided; partly sheathing
PLANT — prickly or hairy
STEM — may be square

VALERIAN, *Valerianaceae* *

FLOWERS — small, white; in flat-top or domed cluster
SEED — 1, with white parachute
LEAVES — opposite; often feather-lobed
STEM — may be square

SEED CAPSULE INSIDE FLOWER

GENTIAN, *Gentianaceae* 80–81
- PETAL LOBES — 4-5
- FLOWERS — erect; solitary or grouped
- SEED CAPSULE — long, plump
- LEAVES — untoothed; opposite (exc. Bogbean — 3-divided)

PHLOX, *Polemoniaceae* 82–83
- PETAL LOBES — often turn back to form flat face above slender tube
- SEED CAPSULE — few-seeded
- LEAVES — opposite (exc. Jacob's Ladder — feather-divided)

VERBENA, *Verbenaceae* 82–83
- PETAL LOBES — as in Phlox
- FLOWERS — small, in spikes
- SEEDS — 4
- LEAVES — opposite
- STEM — square

PRIMROSE (in part), *Primulaceae* 84–85
- PETAL LOBES — as in Phlox
- FLOWERS — in umbels
- LEAVES — in basal rosette

BORAGE, *Boraginaceae* 86–87
- PETAL LOBES — some as in Phlox
- FLOWERS — on uncurling spikes
- SEEDS — 4; hard; occ. bur-like
- PLANT — usually hairy or bristly

PLANTAIN, *Plantaginaceae* 88
- PETAL LOBES — 4
- FLOWERS — minute; papery; crowded on leafless spike
- LEAVES — parallel-veined; in basal rosette; tough fibered

DOGBANE, *Apocynaceae* 89
- FRUIT — slender pod
- SEED — with parachute
- LEAVES — opposite; untoothed
- PLANT — has milky juice

WATERLEAF, *Hydrophyllaceae* 88–89
- STAMENS — often extend beyond tube
- SEED CAPSULE — few-seeded
- LEAVES — lobed or feather-divided
- PLANT — some hairy

TOMATO, *Solanaceae* 89–91
- LEAVES — may be lobed or divided
- FRUIT — berry or dry capsule
- PLANT — erect; occ. a vine or trailer; occ. rank-scented; occ. spiny

MORNING GLORY, *Convolvulaceae* 90–91
- FLOWERS — funnel-shaped
- PLANT — mostly twining, non-woody vines; occ. with milky juice

TRUMPET CREEPER, *Bignoniaceae* *
- FLOWERS — trumpet-shaped
- LEAVES — opposite, feather - divided
- FRUIT large long pod
- PLANT — woody vine, climbing by rootlets

NOTES

4 equal petal lobes may be found in a few members of the Mint and Snapdragon families.

Tubular-appearing flowers with joined sepals enclosing separate petals occur in the Mustard and Pink families.

Petals or petal-like sepals may be slightly joined at base in Rose, Saxifrage, Smartweed, Sandalwood, Mallow, and Wood-Sorrel families; and in Loosestrifes and Shooting Stars of Primrose family.

* not illustrated

Type family:
MINT, *Labiatae* 92—97

PETAL LOBES — 4-5, unequal, forming a 2-lipped flower
SEPAL LOBES — 5, often unequal
FLOWERS — variously arranged, but often in rings just above leaf-stem junctures
SEED CAPSULE — inside flower; 4-seeded
LEAVES — opposite; undivided; may be toothed or lobed
STEM — square
PLANT — usually more or less hairy; often, but not always, aromatic

ASSOCIATED FAMILIES — not aromatic

stem square ONLY in Lopseed, Acanthus, and some in Snapdragon

ACANTHUS, *Acanthaceae* 92
FLOWER — bilaterally to nearly radially symmetrical; many small leaves behind
SEED CAPSULE — 2-celled, many-seeded

LOPSEED, *Phrymaceae* 96
FLOWERS — small, in leafless spikes
FRUIT — 3-hooked stick-tights; 1-seeded

SNAPDRAGON,
Scrophulariaceae 97—101
SEED CAPSULE — 2-celled, many-seeded exc. Cow Wheat
FLOWER — occ. spurred
LEAVES — opposite OR alternate
PLANT — many species NOT hairy

LOBELIA, *Lobeliaceae* 102
PETAL LOBES — sharp-pointed; 2 upper erect, 3 lower spreading
PETAL TUBE — split on upper side
FLOWERS — in racemes
SEED CAPSULES — partly beneath flower; many-seeded
LEAVES — alternate

BLADDERWORT,
Lentibulariaceae 102
FLOWERS — spurred; solitary or in leafless raceme
LEAVES — finely divided (under water); in basal rosette, or often alternate
SEED CAPSULE — many-seeded

BROOMRAPE,
Orobanchaceae 103
LEAVES — reduced to scales; alternate
PLANT — non-green - brown to yellow-white
SEED CAPSULE — many-seeded

NOTES

Other STRONGLY bilateral flowers are found in some Honeysuckles, in all families in Chart F, and in part of Chart A.

SLIGHTLY bilateral tubular flowers may be found in the Plantain, Phlox, Mustard, Trumpet Creeper, Verbena, Honeysuckle, Waterleaf, Blueberry, Tomato families.

SQUARE stems and OPPOSITE leaves occur in some members of the Nettle and Daisy families, and in some in Chart I.

TUBULAR FLOWERS *(in dense heads)*

DAISY, *Compositae* 104–127

FLOWERS — tubes of 2 kinds, crowded into a blossom-like head:
1. DISC — radically symmetrical; 5 petal lobes; often small, many
2. RAY — bilaterally symmetrical; tube flattened, strap-shaped, looking like a petal

SEED — 1; beneath each individual flower; often airborne by a silky parachute formed from the sepals

HEAD — backed by a group of leaves looking like green sepals

LEAVES — alternate, occ. opposite; usually toothed, lobed, or divided

---

DANDELION tribe 104–106
FLOWERS — RAY type *only;* yellow to orange, occ. blue to white

SEED — with parachute exc. Chicory

PLANT — with *milky juice*

THISTLE–LIKE tribes 107–111
(Thistle, Joe-Pye Weed, Ironweed)

FLOWERS — DISC type *only;* appear *fuzzy* with numerous prolonged petal lobes or pistils; purple, pink, white

SEED — with parachute exc. Burdock and Centaureas

DAISY tribe 111–112
HEADS — flat exc. Pineapple Weed

FLOWERS — RAYS: white, when present
— DISCS: yellow, white, green

SEED — NO parachute

LEAVES — *much divided;* strong-scented exc. Daisy, some Wormwoods

---

SUNFLOWER tribe 111, 112–119
(and part of Inula tribe)

HEADS — domed or flat; *over* 1″ wide

FLOWERS — RAYS: conspicuous; yellow, occ. purple, white, pink; occ. none
— DISCS: yellow or dark

SEED — NO parachute exc. Elecampane

LEAVES — *often opposite,* usually undivided

ASTER–LIKE tribes 119–127
(Aster, Ragwort, Inula)

HEADS — *less* than 1″ wide; often very numerous

FLOWERS — RAYS: purple, blue, pink, white, yellow; occ. lacking
— DISCS: yellow, pink, purple, white

SEED — with parachute exc. Gumweed

LEAVES — undivided, occ. lobed

CAPTIONS

The description for each plant illustrated includes: English and botanical name; flowering period; plant size possibilities; habitat, based on plant-communities drawing, p. **xii**; field clues*; occasional mention of an unpictured similar species, with helpful differences noted. (Compass bearings are always *within* our geographical range.)

*Growth habit, for example, is frequently useful to distinguish one related species from another.

SINGLY—the entire plant consists of a *single* stem or two, rising from the root. If the stem branches, it does so ABOVE GROUND (as Milkwort, p. 53).

CLUMP—from a single root rise *many* stems spreading from one point AT GROUND LEVEL, all close to one another (as Seneca Snakeroot, p. 53)

PATCH—the stems branch BELOW GROUND LEVEL, forming a horizontal underground network, from which rise many single erect stems. The whole 'patch' is, in fact, *one* plant (a typical patch-former is Canada Goldenrod, p. 125).

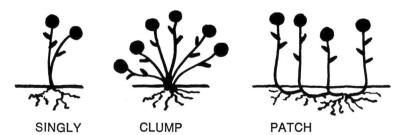

SINGLY   CLUMP   PATCH

TITLES

To show the camera's scale, with the title of each picture is a size of flower, group of flowers, or other character *in* the photograph. It is length (as in Lobelias) or width (as in Trilliums), whichever is greater. The following abbreviations are used: fl.—flower; fl. arr.—flower arrangement; hd.—head, as in Daisy family; col.—column (of flowers), in some Goldenrods.

WATER-PLANTAIN   fl. ¼"

ARROWHEAD   fl. 1¾"

1

CAT-TAIL   fl. arr. 12" tall

WATER-PLANTAIN FAMILY
ARROWHEAD, *Sagittaria latifolia* /
July–Aug. / 6"–48" / Shallow water,
mud / Leaves vary from broad to
narrow

WATER-PLANTAIN, *Alisma plant-*
ago-aquatica / May–Sept. / 4"–40"
/ Shallow water, mud
(ARROW ARUM, *Peltandra virginica*—in
Calla family: leaves arrow-shaped, flower
like Jack in the Pulpit)

CAT-TAIL FAMILY
CAT-TAIL, *Typha latifolia* / May–
July / 3'–9' / Shallow water / Leaf
D-shaped in cross section
(NARROWLEAF CAT-TAIL, *T. angustifo-*
lia: flower spike interrupted by 1" of
bare stem)

BUR-REED FAMILY
BUR-REED, *Sparganium eurycarpum*
/ June–July / 1'–3½' / Shallow
water / Leaf triangular in cross section

BUR-REED   fl. arr. ¾"

SPIDERWORT FAMILY

SPIDERWORT, *Tradescantia ohiensis* / April–July / 16"–40" / Inland sands, medium to dry meadows, prairies, roadsides / Blue, pink, white (Other species have hairy sepals)

DAYFLOWER, *Commelina communis* / July–Sept. / 4"–14" / Damp shade, often a garden weed
(*C. erecta:* leaf-sheath flaring at top; dry sand)

SPIDERWORT   fl. 1"

DAYFLOWER   fl. 1"

PIPEWORT FAMILY

PIPEWORT, *Eriocaulon septangulare* / June–Sept. / 1"–8" / Shallow water, boggy shores

PICKEREL WEED FAMILY

PICKEREL WEED, *Pontederia cordata* / June–Aug. / 1'–3½' / Shallow water, mud

FLOWERING RUSH FAMILY

FLOWERING RUSH, *Butomus umbellatus* / June–Aug. / 1'–5' / Shores of Great Lakes, St. Lawrence River / Leaves grass-like

CALLA FAMILY

JACK IN THE PULPIT, *Arisaema triphyllum* / April–June / 1'–3' / Medium to wet woods and forests

GREEN DRAGON, *Arisaema dracontium* / May–June / 1'–3' / Wet woods

SWEET FLAG, *Acorus calamus* / June / 1'–3' / Shallow water / Leaves flat, aromatic

WILD CALLA, *Calla palustris* / June / 5"–10" / Bogs, cold shallow water / Stems creep or float

SKUNK CABBAGE, *Symplocarpus foetidus* / March–May / Wet ground near springs, sun or shade / Leaves to 3' follow flowers

PIPEWORT
fl. arr. ¼"

PICKEREL WEED
fl. arr. 4"

FLOWERING RUSH   fl. 1″

SWEET FLAG   fl. arr. 3″

JACK IN THE PULPIT
fl. arr. 4″ tall

WILD CALLA   fl. arr. 2″

SKUNK CABBAGE   fl. arr. 4½″

GREEN DRAGON
fl. arr. 8″ long

DAYLILY fl. 4"

LILY FAMILY

DAY LILY, *Hemerocallis fulva* / June–Aug. / 2'–4' / Common garden escape to roadsides, forming dense patches

TURK'S CAP LILY, *Lilium superbum* / July–Aug. / 3'–6' / Wet meadows, streambanks / Orange to yellow

WOOD LILY, *L. philadelphicum* / June–Aug. / 1'–3' / Dry woods and forests, meadows, prairies, sandy soil / Orange to yellow

YELLOW FAWN-LILY, *Erythronium americanum* / April–May / 4"–8" / Dry to wet woods and forests (Similar, and in similar habitats, *E. albidum:* white, pale pink)

YELLOW BEADLILY, *Clintonia borealis* / May–June / 6"–12" / Medium to wet forests

PRAIRIE TRILLIUM, *Trillium recurvatum* / April–May / 8"–16" / Medium prairies and woods—west / Flower stalkless; leaves mottled

NODDING TRILLIUM, *T. cernuum* / May–June / 8"–16" / Medium to wet woods and forests

PURPLE TRILLIUM, *T. erectum* / April–May / 8"–16" / Medium woods and forests—east

SNOW TRILLIUM, *T. nivale* / March–April / 3"–6" / Medium limy woods / Leaves stalked

LARGE TRILLIUM, *T. grandiflorum* / May–June / 8"–16" / Dry to medium woods, medium forests / Aging flowers may change to pink

TURK'S CAP
LILY fl. 3"

WOOD LILY fl. 2½"

YELLOW BEADLILY fl. ⅝"

YELLOW FAWN LILY fl. ⅝"

PRAIRIE TRILLIUM
fl. 1"

NODDING TRILLIUM
fl. 1½"

PURPLE TRILLIUM   fl. 1½"

SNOW TRILLIUM   fl. 1"

LARGE TRILLIUM   fl. 2"

INDIAN CUCUMBER ROOT, *Medeola virginiana* / May–July / 1'–2½' / Medium woods and forests

SOLOMON'S SEAL   fl. ½"

STARRY SOLOMON'S PLUME
fl. arr. 2"

3-LEAF SOLOMON'S PLUME   fl. arr. 2"

SOLOMON'S PLUME   fl. arr. 4"

TWISTED STALK   fl. ⅜"

WILD HYACINTH   fl. ½"

ASPARAGUS   fl. ⅜"

LILY FAMILY

SOLOMON'S SEAL, *Polygonatum biflorum* / May–July / 1'–5' / Dry to medium woods and prairies

SOLOMON'S PLUME, *Smilacina racemosa* / May–June / 16"–32" / Dry to medium woods and forests

STARRY SOLOMON'S PLUME, *S. stellata* / May–June / 8"–24" / Wet to dry prairies, woods, inland sands, beaches

3-LEAF SOLOMON'S PLUME, *S. trifolia* / May–June / 4"–16" / Wet forests, bogs

TWISTED STALK, *Streptopus roseus* / June–July / 1'–3' / Medium forests / Flowers attached *below* leaves; stem branched

ASPARAGUS, *Asparagus officinallis* / May–June / 2'–7' / Fields, open woods, fencerows

WILD HYACINTH, *Camassia scilloides* / April–June / 1'–2' / Medium to wet prairies and open woods

BELLWORT, *Uvularia grandiflora* / April–May / 8"–20" / Dry to medium woods, medium forests / Leaf-base clasps stem / Stem branched

WILD OATS, *U. sessilifolia* / April–May / 4"–12" / Medium forests and clearings / Stem branched

COLIC ROOT, *Aletris farinosa* / June–July / 2'–3½' / Sandy soil / Flower surface minutely "bumpy" (FALSE ASPHODEL, *Tofieldia glutinosa*: leaves grasslike; stem sticky; wet limy soil)

WILD LILY OF THE VALLEY, *Maianthemum canadense* / May–June / 4"–6" / Wet to dry woods and forests / Flower 4-petaled

WILD LILY OF THE VALLEY fl. ⅜"

BELLWORT fl. 1⅛"

WILD OATS fl. ⅞"

COLIC ROOT fl. ½"

LILY FAMILY

NODDING WILD ONION, *Allium cernuum* / July–Aug. / 1′–2′ / Prairies, dry hillsides / Pink to white

WILD LEEK, *A. tricoccum* / June–July / 4″–12″ / Dry to wet woods, forests / Flat spring leaves, onion odor, die before flowers bloom

WILD GARLIC, *A. canadense* / May–June / 8″–24″ / Prairies, dry to wet woods / Bulblets among flowers characteristic

FALSE HELLEBORE, *Veratrum viride* / June–July / 3′–6′ / Wet woods and meadows—east
(*V. woodii:* flowers purple-black; woods —south)

DEVIL'S BIT, *Chamaelirium luteum* / June / 1′–4′ / Moist woods, bogs —south

WHITE CAMAS, *Zygadenus elegans* / June–Sept. / 1′–3′ / Dry to wet limy prairies, meadows, beaches, bogs / Large nectar pads on petals
(BUNCHFLOWER, *Melanthium virginicum:* petals stalked, stem hairy; FEATHER FLEECE, *Stenanthium gramineum:* petals slender-pointed, lack nectar spots—South)

YUCCA, *Yucca filamentosa* / July–Sept. / to 10′ / Dry sandy soil, often cultivated

CARRION FLOWER, *Smilax herbacea* / May–June / Climbing by tendrils to 7′ / Moist to dry woods, fencerows
(LOW CARRION FLOWER, *S. ecirrhata:* 1′–3′, not climbing; flowers below leaves)

GREENBRIAR, *S. hispida* / May–June / Climbing by tendrils to 10′ / Wet to medium woods and forests, fencerows / Stem spiny

AMARYLLIS FAMILY

STAR GRASS, *Hypoxis hirsuta* / April–July / 4″–12″ / Dry open woods, dry to wet prairies

NODDING WILD ONION   fl. ¼″

WILD LEEK   leaves

WILD LEEK
fl. arr. 1¼″

FALSE HELLEBORE   fl. 1″

WILD GARLIC   fl. ½"

DEVIL'S BIT   fl. arr. 8"

WHITE CAMAS   fl. ⅜"

YUCCA   fl. 2⅜"

CARRION FLOWER   fl. arr. 1½"

STAR GRASS   fl. ⅝"

GREENBRIAR   fl. arr. 1½"

DWARF LAKE IRIS   fl. 2"

WILD IRIS   fl. 1⅞"

YELLOW FLAG   fl. 3"

BLUE-EYED GRASS   fl. ½"

IRIS FAMILY

DWARF LAKE-IRIS, *Iris lacustris* / May–June / 2"–3" / Limy woods and shores around the Great Lakes

WILD IRIS, *I. shrevei* / May–July / 1'–3' / Wet meadows, streambanks / Conspicuous yellow blotch on falls

YELLOW FLAG, *I. pseudacorus* / April–June / 2'–3' / Shallow water, shores

BLUE-EYED GRASS, *Sisyrinchium campestre* / May–July / 4"–16" / Dry to medium meadows, prairies, and woods / Blue, white, occ. violet

ORCHID FAMILY

SHOWY LADYSLIPPER, *Cypripedium reginae* / June–July / 16'–40' / Wet woods, bogs / Hairy stem and leaves can be a skin irritant

PINK LADYSLIPPER, *C. acaule* / April–June / 8"–16" / Acid soil of bogs, dry woods and forests, cliffs, inland sands

RAM'S HEAD ORCHID, *C. arietinum* / May–June / 4"–16" / Wet forests

YELLOW LADYSLIPPER, *C. pubescens* / May–June / 8"–30" / Dry to wet woods and forests, occ. prairies (SMALL YELLOW LADYSLIPPER, *C. parviflorum:* twisted petals dark brown; bogs, wet forests; WHITE LADYSLIPPER, *C. candidum:* wet shrubby meadows and prairies)

ROUNDLEAF ORCHID, *Habenaria orbiculata* / July–Aug. / 1'–2' / Medium to wet woods and forests

HOOKER'S ORCHID, *H. hookeri* / June–July / 8"–16" / Wet to medium forests, occ. woods / Leaves 2, round, flat on ground

SHOWY LADYSLIPPER   pouch 1¾″

PINK LADYSLIPPER   pouch 1¾″

YELLOW LADYSLIPPER pouch 2½″

RAM'S-HEAD ORCHID pouch ¾″

HOOKER'S ORCHID   lip ⅜″

ROUNDLEAF ORCHID   fl. arr. 6½″

ORCHID FAMILY

CLUB-SPUR ORCHID, *Habenaria clavellata* / July–Aug. / 4″–16″ / Bogs, wet acid soil / Lower leaf slender, blunt

PURPLE FRINGED ORCHID, H. psycodes / June–Aug. / 1′–3′ / Wet meadows, wet open woods and forests, streambanks

PRAIRIE FRINGED ORCHID, H. leucophaea / June–July / 16″–40″ / Wet meadows and prairies, bogs
(RAGGED FRINGED ORCHID, *H. lacera:* flowers greenish-white)

YELLOW FRINGED ORCHID, H. ciliaris / July–Aug. / 16″–32″ / Bogs, acid sand, occ. disturbed ground

FROG ORCHID, H. viridis / June–July / 8″–20″ / Medium to dry woods and forests
(NORTHERN BOG ORCHID, *H. hyperborea:* flowers white to green)

YELLOW TWAYBLADE, Liparis loeselii / June–Aug. / Wet forests and woods, bogs, shores

TWAYBLADE, L. liliifolia / June–July / 6″–8″ / Dry open woods, often abundant among saplings in cleared ground

HEARTLEAF TWAYBLADE, Listera cordata / July / 4″–8″ / Bogs, wet forests

ADDER'S MOUTH, Malaxis uniflora / July–Aug. / 4″–12″ / Bogs, moist forests and clearings

SNAKEMOUTH, Pogonia ophioglossoides / May–July / 8″–12″ / Bogs, mossy shores / Oval leaf on stem

SHOWY ORCHIS, Orchis spectabilis / May–June / 4″–8″ / Medium open woods

CLUB-SPUR ORCHID   fl. arr. 2″

PURPLE FRINGED ORCHID
fl. arr. 7″

YELLOW FRINGED ORCHID   fl. arr. 6″

PRAIRIE FRINGED ORCHID
fl. arr. 6½″

FROG ORCHID fl. arr. 6½"

**HEARTLEAF
TWAYBLADE** fl. ¼"

ADDER'S MOUTH
fl. arr. 4"

YELLOW TWAYBLADE lip ¼"

TWAYBLADE lip ½"

SHOWY ORCHIS
lip ¾"

SNAKEMOUTH
fl. 1½"

THREE BIRDS fl. ¾″

DRAGONMOUTH fl. 2″

CALYPSO lip ¾″

WHORLED POGONIA lip ¾″

SPOTTED CORAL ROOT fl. arr. 6″

GRASS PINK fl. 1½″　　**CORAL ROOT fl. arr. 13″**　　**HELLEBORINE fl. arr. 12″**

ORCHID FAMILY

THREE BIRDS, *Triphora triantho-phora* / Aug.–Sept. / 4"–12" / Wet to medium forests, leafmold

DRAGONMOUTH, *Arethusa bulbosa* / May–June / 4"–12" / Bogs, wet meadows / Grass-like leaf follows flower

GRASS PINK, *Calopogon pulchellus* / June–July / 1'–2' / Bogs, wet sandy soil / Leaf grass-like

CORAL ROOT, *Corallorhiza striata* / May–July / 8"–12" / Wet to medium forests / No leaves
(EARLY CORAL ROOT, *C. trifida:* 3"–10"; pale yellow-green; bogs, wet forests)

SPOTTED CORAL ROOT, *C. maculata* / July–Sept. / 8"–20" / Wet to dry woods and forests / No leaves

CALYPSO, *Calypso bulbosa* / May–June / 2"–6" / Damp mossy forests / 1 small broad leaf at base

WHORLED POGONIA, *Isotria verticillata* / May–June / 8"–16" / Acid soil: bogs, sandy maple woods

HELLEBORINE, *Epipactis helleborine* / July–Aug. / 1'–3' / Clayey forests and woods, woody dunes

LADIES' TRESSES, *Spiranthes cernua* / July–Sept. / 8"–36" / Bogs, damp to dry sandy soil, wet to dry prairies

SLENDER LADIES' TRESSES, *S. gracilis* / July–Aug. / 4"–24" / Damp to dry sandy soil and woods openings / Leaves die by blooming time

ADAM AND EVE, *Aplectrum hyemale* / May–June / 1'–2' / Medium woods / Pleated leaf appears after blooming, remains into spring

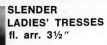

SLENDER
LADIES' TRESSES
fl. arr. 3½"

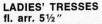

LADIES' TRESSES
fl. arr. 5½"

ADAM AND EVE   fl. ½"

ADAM AND EVE   leaf

RATTLESNAKE PLANTAIN
fl. arr. 3½"

WILD YAM fl. arr. 3½"

CLAMMY WEED fl. ⅛"

ORCHID FAMILY

RATTLESNAKE PLANTAIN, *Goodyera pubescens* / July–Aug. / 8"–16" / Dry woods and forests
(Other RATTLESNAKE PLANTAINS vary in leaf markings)

YAM FAMILY

WILD YAM, *Dioscorea villosa* / May–Aug. / Twining to 18' / Wet to medium woods / Female flowers form 3-winged seed capsule; male flowers tiny, numerous, white

CAPER FAMILY

CLAMMY WEED, *Polanisia dodecandra* / July–Sept. / 8"–20" / Dry disturbed ground, sand, gravel / Stem sticky; leaf 3-divided

MUSTARD FAMILY

BLACK MUSTARD, *Brassica nigra* / June–Sept. / 6"–60" / Disturbed ground, fields / Leaves coarse, bristly

YELLOW CRESS, *Rorippa islandica* / May–Oct. / 18"–40" / Wet disturbed ground, shores / Pods oval, plump

TUMBLE MUSTARD, *Sisymbrium altissimum* / June–July / 6"–40" / Disturbed ground / Flowers very pale yellow
(*S. officinale:* leaf-divisions *triangular;* TANSY MUSTARD, *Descurainia pinnata:* feathery leaves *twice* divided)

WORMSEED MUSTARD, *Erysimum cheiranthoides* / June–Aug. / 8"–40" / Disturbed ground

WINTER CRESS, *Barbarea vulgaris* / April–June / 8"–32" / Fields, disturbed ground, often moist / Divisions of lower leaves nearly round

HOARY ALYSSUM, *Berteroa incana* / May–Sept. / 10"–28" / Disturbed ground / Pods oval, plump; plant gray-hairy

SHEPHERD'S PURSE, *Capsella bursa-pastoris* / April–Aug. / 4"–24" / Disturbed ground / Pods heart-shaped, flat

WATERCRESS, *Nasturtium officinale* / May–Oct. / 4"–18" / Clear streams and springs, usually limy

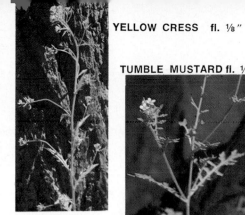

YELLOW CRESS   fl. ⅛″

TUMBLE MUSTARD fl. ⅛″

BLACK MUSTARD   fl. ¼″

WORMSEED MUSTARD   fl. ⅛″

WINTER CRESS   fl. ¼″

HOARY ALYSSUM
fl. ⅛″

SHEPHERD'S PURSE   fl. ⅛″

WATERCRESS   fl. ⅛″

PENNY CRESS fl. ⅜"

ROCK CRESS fl. ⅜"

TOOTHWORT fl. ⅜"

DRABA fl. ⅛"

SICKLEPOD fl. ⅛"

PEPPER GRASS fl. ⅛"

CRINKLEROOT fl. ⅜"

MUSTARD FAMILY

PENNY CRESS, *Thlaspi arvense* /
/ April–June / 4″–20″ / Disturbed
ground / Pods flat, round, ½″

ROCK CRESS, *Arabis lyrata* / May–
July / 4″–16″ / Dry sandy or rocky
soil, sun or shade / Pods very slender; basal leaves often feather-
divided

SICKLEPOD, *A. canadensis* / May–
July / 1′–3′ / Dry to medium woods,
rocky places / Pods slender, mature
ones hang down
(TOWER MUSTARDS, with leaf-bases
around stem: *A. glabra,* pods erect; *A.
laevigata,* pods curving outward)

DRABA, *Draba reptans* / April–May
/ 2″–10″ / Disturbed ground, sandy
or rocky prairies / Pods oval; leaves
tiny, bristly
(Other DRABAS may have toothed
leaves)

TOOTHWORT, *Dentaria laciniata;*
CRINKLEROOT, *D. diphylla* / April–
May / 8″–16″ / Medium to wet for-
ests and woods

PEPPER GRASS, *Lepidium campestre* / May–June / 6″–30″ / Disturbed
ground / Pods flat or concave,
round, ¼″
(*L. virginicum:* pods flat, round, 1/8″)

HORSERADISH, *Armoracia rusticana*
/ July / 2′–3′ / Moist disturbed
ground / Lower leaves large, coarse

BITTER CRESS, *Cardamine pensylvanica* / April–July / 6″–24″ / Wet
woods, streambanks

CARDAMINE, *C. bulbosa* / April–
June / 10″–40″ / Wet woods and
meadows, shallow water / Leaves
undivided, lower ones nearly round
(WOODLAND CARDAMINE, *C. douglas-
sii:* flowers pink)

CUCKOO FLOWER, *C. pratensis* /
April–July / 8″–20″ / Bogs, moist
open places / Leaves feather-divided; flowers white, pink

HORSERADISH fl. ¼″

BITTER CRESS fl. ⅛″

**CARDAMINE
fl. ¼″**

CUCKOO FLOWER fl. ¾″

SEA ROCKET fl. ¼"

SWEET ROCKET fl. ¾"

TWINLEAF fl. 1¼"

MAYAPPLE fl. 1¾"

BLUE COHOSH fl. ⅜"

BLOODROOT fl. 1¾"

CELANDINE fl. ⅝"

MEADOW BEAUTY   fl. ¾″

GOLDEN FUMEWORT   fl. ½″

PALE CORYDALIS   fl. ⅝″

MUSTARD FAMILY

SWEET ROCKET, Hesperis matronalis / May–Sept. / 2′–3′ / Garden escape to disturbed ground, sun or shade / White to purple, fragrant; leaves not divided
(PURPLE ROCKET, Iodanthus pinnatifidus: without hairs; leaves stalked; riverbanks)

SEA ROCKET, Cakile edentula / July–Oct. / 4″–20″ / Great Lakes shores / Leaves fleshy; pod plump

BARBERRY FAMILY

TWINLEAF, Jeffersonia diphylla / April–May / 4″–8″ / Medium woods / Seed pod has opening lid

MAYAPPLE, Podophyllum peltatum / May / 1′–2′ / Dry to medium woods / Ripe fruit yellowish

BLUE COHOSH, Caulophyllum thalictroides / April–May / 1′–3′ / Medium to wet woods and forests / Flower 6-petaled, green or brownish

POPPY FAMILY

BLOODROOT, Sanguinaria canadensis / April–May / 3″–6″ / Dry to medium woods and forests / Pod slender, splits lengthwise

CELANDINE, Chelidonium majus / April–Sept. / 1′–3′ / Medium to wet woods / Pod slender
(CELANDINE POPPY, Stylophorum diphyllum: flowers 1½″; pods plump, white-hairy)

MEADOW BEAUTY FAMILY

MEADOW BEAUTY, Rhexia virginica / July–Sept. / 8″–24″ / Moist sandy meadows

BLEEDING HEART FAMILY

GOLDEN FUMEWORT, Corydalis aurea / April–June / 3″–16″ / Disturbed ground in sandy or rocky woods and forests

PALE CORYDALIS, C. sempervirens / May–Sept. / 1′–3′ / Dry or rocky woods and forests, cliffs

DUTCHMAN'S BREECHES   fl. ⅝"

SQUIRREL CORN   fl. ½"

MOUNTAIN FRINGE   fl. ⅝"

FUMITORY   fl. ¼"

ROCK SANDWORT   fl. ¼"

STARRY CHICKWEED   fl. ¼"

SANDWORT
fl. ¼"

BLEEDING HEART FAMILY

DUTCHMAN'S BREECHES, *Dicentra cucullaria* / April–May / 4″–12″ / Dry to wet woods, medium forests, cliffs

SQUIRREL CORN, *D. canadensis* / April–May / 4″–12″ / Medium forests

MOUNTAIN FRINGE, *Adlumia fungosa* / June–Sept. / Climbing by twisting leaf stalks to 10′ / Wet forest openings, rocky slopes

FUMITORY, *Fumaria officinalis* / May–Sept. / Climbing or trailing 8″–24″ / Disturbed ground of woods and forests

PINK FAMILY

SANDWORT, *Arenaria lateriflora* / May–Aug. / 4″–12″ / Dry to medium woods and forests
(The SPURREYS, with numerous tiny slender leaves, flowers 1/8″: FIELD, *Spergularia arvensis,* white, leaves whorled; SAND, *S. rubra,* pink, leaves white-appendaged)

ROCK SANDWORT, *A. stricta* / June–Sept. / 4″–16″, in moss-like cushions to 6″ diam. / Cliffs, gravel, dry prairies
(PEARLWORT, *Sagina nodosa,* fl. 1/8″, moist rocky or sandy soil—north)

STARRY CHICKWEED, *Cerastium arvense* / May–Aug. / 6″–16″ / Disturbed ground: sand, gravel, rocks

MOUSE-EAR CHICKWEED, *C. vulgatum* / May–Oct. / 1″–12″ / Disturbed ground, sun or shade / Leaves hairy; common in lawns, pastures

GIANT CHICKWEED, *Stellaria aquatica* / May–Oct. / 2″–24″ / Wet to damp disturbed ground, woods
(A very common lawn CHICKWEED, *S. media:* fl. 1/8″)

STARRY CAMPION, *Silene stellata* / June–Aug. / 1′–3′ / Medium woods and clearings / Lower leaves 4-whorled
(*S. nivea:* petals not fringed, all leaves opposite)

STARWORT, *Stellaria longifolia* / May–Aug. / 4″–20″ / Wet meadows, damp woods and forests

MOUSE-EAR CHICKWEED fl. ⅛″

GIANT CHICKWEED fl. ¼″

STARRY CAMPION fl. ⅝″

STARWORT fl. ⅛″

BLADDER CAMPION fl. ½″

GARDEN CATCHFLY
fl. ½″

SLEEPY CATCHFLY fl. ⅛″

WHITE CAMPION fl. ¾″

FIRE PINK fl. ¾″

FORKED CATCHFLY fl. ⅝″

BOUNCING BET   fl. ¾″

CORN COCKLE   fl. 2½″

DEPTFORD PINK   fl. ⅜″

PINK FAMILY

WHITE CAMPION, *Lychnis alba* / June–Oct. / 12″–40″ / Disturbed ground / Flowers on male plant smaller, less inflated

BLADDER CAMPION, *Silene cucubalus* / June–Oct. / 8″–32″ / Disturbed ground / Leaves blue-green

GARDEN CATCHFLY, *S. armeria* / June–July / 4″–28″ / Disturbed ground

SLEEPY CATCHFLY, *S. antirrhina* / June–Sept. / 2″–32″ / Dry disturbed ground, rocks / Stem has sticky sections

FORKED CATCHFLY, *S. dichotoma* / June–Oct. / 1′–3′ / Disturbed ground / Flowers all face one way on forked stem

FIRE PINK, *S. virginica* / May–Sept. / 8″–32″ / Medium woods—open or rocky—south, east

BOUNCING BET, *Saponaria officinalis* / July–Oct. / 1′–3′ / Disturbed ground / White or pink; forms dense patches

CORN COCKLE, *Agrostemma githago* / July–Sept. / 2′–3′ / Disturbed ground / Sepals longer than petals

DEPTFORD PINK, *Dianthus armeria* / July–Aug. / 8″–24″ / Dry disturbed ground / Petals white-spotted
(COW HERB, *Vaccaria segetalis:* fl. 1/4″, not spotted; sepals have sharp vertical ridge)

ST. JOHN'S-WORT FAMILY

MARSH ST. JOHN'S-WORT, *Triadenum virginicum* / July–Aug. / 12"–18" / Bogs, shallow water / Pink

GREAT ST. JOHN'S-WORT, *Hypericum pyramidatum* / July–Aug. / 2'–5' / Wet to medium woods and forests, openings, streambanks

SHRUBBY ST. JOHN'S-WORT, *H. kalmianum* / July–Aug. / 6"–40" / Moist sand or rocks, chiefly near the Great Lakes / Woody, to 6' diam.

ST. JOHN'S-WORT, *H. perforatum* / June–Sept. / 2'–3' / Disturbed ground, fields, roadsides
(In wetlands: *H. ellipticum:* with runners, fl. 3/8"; *H. majus:* solitary)

DOTTED ST. JOHN'S-WORT, *H. punctatum* / June–Aug. / 2'–3' / Woods and forest openings and edges / Flowers brown-streaked

ORANGE GRASS, *H. gentianoides* / July–Oct. / 4"–24" / Sand or rocks, full sun / Leaves merely tiny scales

MARSH ST. JOHN'S-WORT fl. ⅜"
GREAT ST. JOHN'S-WORT fl. 2"

PURPLE LOOSESTRIFE FAMILY

PURPLE LOOSESTRIFE, *Lythrum salicaria* / July–Sept. / 1'–4' / Wet meadows, shores, shallow water / Leaves chiefly opposite
(WINGED PURPLE LOOSESTRIFE, *L. alatum:* flowers open a few at a time; upper leaves mostly alternate)

WATER WILLOW, *Decodon verticillatus* / Aug.–Sept. / Woody arching branches 1'–9' / Shallow water

MILKWEED FAMILY

WHITE MILKWEED, *Asclepias verticillata* / July–Sept. / 8"–20" / Dry prairies, fields, roadsides, sandy soil / Leaves whorled; forms patches

GREEN MILKWEED, *A. hirtella* / June–Aug. / 16"–40" / Dry sandy soil and prairies
(*A. viridiflora:* flower-umbels not stalked; leaves narrow-oval, few)

SMALL GREEN MILKWEED, *A. lanuginosa* / June / 4"–12" / Dry woods, prairies / Plant hairy

POKE MILKWEED, *A. exaltata* / June–July / 3'–6' / Dry to medium woods

SHRUBBY ST. JOHN'S-WORT
fl. 1¼"

WATER WILLOW fl. ½"

ST. JOHN'S-WORT   fl. 5/8″
ORANGE GRASS   fl. 1/8″

DOTTED ST. JOHN'S-WORT   fl. 1/4″

WHITE MILKWEED
fl. 1/4″

GREEN MILKWEED   fl. 1/4″

PURPLE
LOOSESTRIFE   fl. 1/2″

SMALL GREEN MILKWEED
fl. 1/4″

POKE MILKWEED   fl. 1/2″

BUTTERFLY WEED   fl. ⅜″

MILKWEED   fl. ⅝″

MARSH MILKWEED   fl. ⅜″

PURSLANE   fl. ⅛″

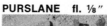

FAME FLOWER
fl. ¼″

BLUNTLEAF MILKWEED   fl. ⅝″

SPRING BEAUTY   fl. ½″

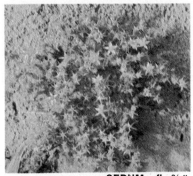

SEDUM fl. ⅜"

3-LEAF SEDUM fl. ⅜"

DITCH STONECROP fl. ¼"

MILKWEED FAMILY

BUTTERFLY WEED, *Asclepias tuberosa* / July–Sept. / 1'–3' / Inland sands, dry prairies and open woods / Leaves alternate; juice not milky

MARSH MIILKWEED, *A. incarnata* / July–Aug. / 1'–4' / Wet prairies, marshes, shores, streambanks

MILKWEED, *A. syriaca* / June–Aug. / 1½'–6' / Meadows, fields, prairies / Cream to pink; pods usually warty; leaves fine-hairy beneath
(*A. sullivantii:* leaves not hairy, strong midrib; wet to medium prairies. PURPLE MILKWEED, *A. purpurascens:* pink to purple; flowers held above leaves; dry prairies, open woods)

BLUNTLEAF MILKWEED, *A. amplexicaulis* / June–July / 1'–3' / Dry prairies and woods openings, sandy soil / Leaf wavy, with pink midrib

PURSLANE FAMILY

PURSLANE, *Portulaca oleracea* / June–Sept. / 1"–20", forming mats to 20" diam. / Disturbed ground / Leaves thick, fleshy

FAME FLOWER, *Talinum rugospermum* / July–Aug. / 4"–8" / Cliffs, sandy soil / Flowers open late afternoon

SPRING BEAUTY, *Claytonia virginica* / April–June / 3"–6" / Wet to medium woods and forests / Petals often striped
(*C. caroliniana,* similar in form and habitat, has broader leaves)

SEDUM FAMILY

SEDUM, *Sedum acre* / June–July / Dense mats / Garden escape to sandy or rocky soil / Leaves tiny, close to ground (Other Sedums also escape)

3-LEAF SEDUM, *S. ternatum* / May–June / Creeper, to 8" high / Rocks, cliffs, woods—east

DITCH STONECROP, *Penthorum sedoides* / July–Sept. / 6"–28" / Muddy shores, streambanks

BUTTERCUP FAMILY

GOLDEN SEAL, *Hydrastis canadensis* / May / 6"–20" / Medium woods / No petals

WHITE BANEBERRY, *Actaea alba* / May–June / 1½'–3' / Medium woods and forests / White berries on *thick* red stalks
(RED BANEBERRY, *A. rubra:* flowers and leaves similar; red or white berries on *thin* stalks)

VIRGIN'S BOWER, *Clematis virginiana* / July–Aug. / Climbing or arching to 9' / Woods and forest clearings and edges, streambanks

ROCK CLEMATIS, *C. verticillaris* / May–June / Climbing to 6' / Medium rocky woods and forests / Clematis leaves 3-divided, opposite

BUGBANE, *Cimicifuga racemosa* / July–Aug. / 3'–7' / Dry to medium woods—east / No petals

MONKSHOOD, *Aconitum uncinatum* / Aug.–Oct. / 1½'–2½' / Wet woods, cliffs / Flower helmet-like

COLUMBINE, *Aquilegia canadensis* / May–July / 1'–3' / Dry open woods and forests, cliffs / Flower spurred; leaflets longer than broad

LARKSPUR, *Delphinium tricorne* / April–May / 8"–24" Medium to moist woods / Flower spurred; blue, white, violet
(*D. virescens:* June–July; dry prairies)

MEADOWRUE, *Thalictrum dasycarpum* / June–July / 3'–5' / Moist meadows, streambanks / Neither Meadowrue has petals

EARLY MEADOWRUE, *T. dioicum* / April–May / 8"–28" / Dry to wet woods, medium forests / Leaflets short, rounded in both species

GOLDTHREAD, *Coptis trifolia* / May–June / 2"–3" / Wet to medium forests, bogs / Underground stems yellow

GOLDEN SEAL   fl. ⅜"

WHITE BANEBERRY   fl. arr. 2"

VIRGIN'S BOWER   fl. ⅜"

ROCK CLEMATIS   fl. 1¾"

BUGBANE fl. arr. 8″

MONKSHOOD fl. ¾″

COLUMBINE
fl. 1¼″

LARKSPUR
fl. ⅝″

EARLY MEADOWRUE fl. arr. 6″

MEADOWRUE
fl. arr. 10″

GOLDTHREAD fl. ⅜″

TALL BUTTERCUP fl. ¾″

PRAIRIE BUTTERCUP fl. ⅜″

TUFTED BUTTERCUP fl. ½″

HOOKED BUTTERCUP fl. ¼″

SWAMP BUTTERCUP fl. 1″

SMALLFLOWER BUTTERCUP fl. ¼″

MARSH MARIGOLD fl. 1″

RUE ANEMONE fl. ⅝"

FALSE RUE ANEMONE fl. ½"

WHITE WATER-CROWFOOT
fl. ⅜"
YELLOW WATER-CROWFOOT

BUTTERCUP FAMILY

TALL BUTTERCUP, *Ranunculus acris* / May–Oct. / 10"–40" / Disturbed or wet ground / Stems tall, few-leaved

PRAIRIE BUTTERCUP, *R. rhomboideus* / April–May / 2"–8" / Dry prairies, open woods / Plant finely-hairy

TUFTED BUTTERCUP, *R. fascicularis* / April–May / 5"–10" / Dry open woods and prairies
(CREEPING CROWFOOT, *R. repens:* meadows, lawns, through July)

HOOKED BUTTERCUP, *R. recurvatus* / May–June / 8"–28" / Medium to wet woods / Basal leaves maple-like

SMALLFLOWER BUTTERCUP, *R. abortivus* / April–June / 8–"20" / Medium to dry woods, cliffs, disturbed ground / Upper leaves slender, lower leaves round
(In marshes, May–Aug.: CURSED CROWFOOT, *R. sceleratus,* similarly smooth but leafier; *R. pensylvanicus* with hairy, more divided leaves)

SWAMP BUTTERCUP, *R. septentrionalis* / May–June / 5"–10" / Wet woods, forests, and meadows / Forms patches

MARSH MARIGOLD, *Caltha palustris* / April–May / 8"–24" / Wet meadows, woods, and forests; marshes, streambanks
(GLOBE FLOWER, *Trollius laxus:* leaves divided—east)

RUE ANEMONE, *Anemonella thalictroides* / April–May / 4"–8" / Dry to medium woods, woods edges / Flowers in umbels, white or pink

FALSE RUE ANEMONE, *Isopyrum biternatum* / April–May / 4"– 16" / Wet to medium woods, forests / Flowers borne singly

YELLOW WATER-CROWFOOT, *Ranunculus flabellaris* / April–June; **WHITE WATER-CROWFOOT, *R. longirostris*** / June– Aug. / Quiet shallow water / Flowers held above water; leaves submerged, finely-divided
(SEASIDE CROWFOOT, *R. cymbalaria:* round-leaved creeper, shores; SPEARWORT, *R. reptans:* slender leaves, damp open ground)

CANADA ANEMONE   fl. ⅝″

WOOD ANEMONE   fl. ⅜″

THIMBLEWEED   fl. ½″

BUTTERCUP FAMILY

CANADA ANEMONE, *Anemone canadensis* / May–Aug. / 8″–36″ / Wet meadows and prairies, shores / Forms patches

WOOD ANEMONE, A. *quinquefolia* / April–June / 2″–8″ / Dry to medium woods and forests / Flowers solitary, white or pink

THIMBLEWEED, A. *cylindrica* / June –Aug. / 1′–3′ / Dry open woods, prairies, inland sands / Plant densely gray-woolly
(*A. virginiana:* plant somewhat hairy; flowers white to green

HEPATICA, *Hepatica acutiloba* / April–May / 2″–6″ / Dry to medium woods and forests / Sepals petal-like, about 8, white, blue, pink; 3 small green leaves behind each flower
(*H. americana:* leaf-lobes rounded. CAROLINA ANEMONE, *A. caroliniana:* flower similar; leaves finely divided; prairies, inland sands)

PASQUE FLOWER, *Anemone patens* / Mar.–May / 4″–16″ / Dry hillsides, prairies, cliffs, open woods / Finely divided, hairy leaves follow flowers

FLAX FAMILY

YELLOW FLAX, *Linum sulcatum* / June–July / 8″–32″ / Dry open woods and prairies, inland sands / Petals fall by noon
(BLUE FLAXES include *L. lewisii:* arching flower stalks, prairies and rocks; *L. usitatissimum:* flower stalks erect, disturbed ground)

WATERLILY FAMILY

WATER SHIELD, *Brasenia schreberi* / June–Aug. / Flower held slightly above water / Leaf slippery beneath

YELLOW WATERLILY, *Nuphar variegatum* / June–Aug. / Flower held 2″–6″ above water / Leaves chiefly feather-veined

WATERLILY, *Nymphaea odorata* / July–Sept. / Flower floats on water / Leaves chiefly palmately-veined

AMERICAN LOTUS, *Nelumbo lutea* / July–Sept. / Flower held 1½′–3′ above water / Leaves round, not notched

PASQUE FLOWER fl. 1¾"

PASQUE FLOWER leaves

HEPATICA fl. ½"

YELLOW FLAX
fl. ⅜"

WATER SHIELD fl. 1"

YELLOW WATERLILY fl. 2"

WATERLILY fl. 5½"

LOTUS leaf

AMERICAN LOTUS
fl. 4"

ROCKROSE FAMILY

PINWEED, *Lechea intermedia* / July
–Sept. / 6"–24" / Inland sands,
open dry woods, rocks

FALSE HEATHER, *Hudsonia tomentosa* / May–July / 1"–24" /
Beaches, inland sands / Evergreen
bushy mats to 4' diam.

FROSTWEED, *Helianthemum canadense* / May–June / 8"–24" / Dry
prairies and open woods, inland
sands
(*H. bicknellii:* plant gray-green; first
flowers clustered, not solitary)

CACTUS FAMILY

PRICKLY PEAR, *Opuntia compressa*
/ June–July / 2"–8" / Dry prairies,
rocks, inland sands / Patches to
6" diam.

ROSE FAMILY

WILD ROSE, *Rosa carolina* / June–
July / ½'–5' / Dry woods and
prairies, inland sands, roadsides /
Pink, white
(*R. blanda:* low, nearly thornless;
SWAMP ROSE, *R. palustris:* tall, nearly
thornless, wet shores; PRAIRIE ROSE,
R. setigera (pink) and MULTIFLORA, *R.
multiflora* (white) have wide-arching
canes and curved thorns.)

THIMBLEBERRY, *Rubus parviflorus*
/ June–July / 18"–40" / Medium
to wet forests
(PURPLE THIMBLEBERRY, *R. odoratus:* flowers rose-purple; woods southeast) Both species are unarmed

DEWBERRY, *R. flagellaris* / May–
June / Ground trailing, to 8" high /
Wet to dry woods and forests, bogs,
sandy and rocky soil / Stems occ.
spiny

STRAWBERRY, *Fragaria virginiana*
/ April–June / 4"–6" / Meadows,
fields, woods and forest edges

BARREN STRAWBERRY, *Waldsteinia fragarioides* / April–May / 4"
–6" / Forests and clearings

STEEPLE BUSH, *Spiraea tomentosa*
/ July–Sept. / 2'–3½' / Meadows,
streambanks, bogs, acid sand

MEADOW SWEET, *S. alba* / June–
Aug. / 3'–6' / Damp meadows,
streambanks

PINWEED
fl. arr. 6"

FALSE HEATHER   fl. ⅛"

FROSTWEED   fl. 1⅜"

PRICKLY PEAR   fl. 3"

WILD ROSE   fl. 1½″

THIMBLEBERRY   fl. 1″

DEWBERRY   fl. ¾″

STRAWBERRY   fl. ½″

BARREN STRAWBERRY   fl. ½″

**STEEPLE BUSH
fl. arr. 5″**

MEADOW SWEET   fl. arr. 5″

QUEEN OF THE PRAIRIE
fl. arr. 6"

GOAT'S BEARD   fl. 1/16"

ROSE FAMILY

QUEEN OF THE PRAIRIE, *Filipendula rubra* / June–July / 3'–7' / Wet meadows and prairies

GOAT'S BEARD, *Aruncus dioicus* / May–June / 3'–6' / Medium woods (For a very different 'Goat's Beard' see Meadow Goat's Beard)

INDIAN PHYSIC, *Gillenia trifoliata* / May–July / 10"–40" / Dry to moist woods—south

CINQUEFOIL, *Potentilla norvegioa* / June–Aug. / 1'–3' / Disturbed ground, wet or dry, sun or shade /Leaf 3-divided

WINELEAF CINQUEFOIL, *P. tridentata* / June–Aug. / 4"–12" / Rock or gravel shores, open forests / Leaf 3–divided, each leaflet 3–toothed

TALL CINQUEFOIL, *P. arguta* / June–July / 12"–40" / Dry woods and prairies / Lower leaves feather-divided

PURPLE CINQUEFOIL, *P. palustris* / June–Aug. / 8"–24" / Bogs, wet meadows, streambanks / Leaf feather-divided, blue-green

SILVERWEED, *P. anserina* / May–Sept. / 2"–8", spreading by slender runners / Beaches, inland sands / Leaf feather-divided, white beneath

SILVERY CINQUEFOIL, *P. argentea* / June–Sept. / 1"–20" / Dry disturbed ground / Leaf 5-divided, whitened beneath

SULPHUR CINQUEFOIL, *P. recta* / June–Aug. / 16"–32" / Disturbed ground / Leaf 7-divided, palmate

INDIAN PHYSIC   fl. ⅞"

TALL CINQUEFOIL fl. ⅜"

CINQUEFOIL fl. ⅜"

**WINELEAF
CINQUEFOIL** fl. ⅜"

PURPLE CINQUEFOIL
fl. ¾"

SILVERWEED fl. ¾"

SILVERY CINQUEFOIL fl. ¼"

SULPHUR CINQUEFOIL fl. ¾"

SHRUBBY CINQUEFOIL
fl. ¾″

OLD-FIELD CINQUEFOIL
fl. ⅜″

WHITE AVENS   fl. ½″

AVENS   fl. ½″

PRAIRIE SMOKE   fl. 1″

DEWDROPS   fl. ½″

PURPLE AVENS   fl. ½″

AGRIMONY fl. ¼ "

VIOLET WOOD-SORREL
fl. ⅜ "

YELLOW WOOD-SORREL fl. ¼ "

WOOD-SORREL fl. ⅜ "

ROSE FAMILY

SHRUBBY CINQUEFOIL, *Potentilla fruticosa* / June–Oct. / 1′–3′ / Bogs, wet limy meadows, shores / Leaf tiny, feather-divided in 5's

OLD-FIELD CINQUEFOIL, *P. simplex* /April–June / 2″–8″, spreading by runners to 10′ diam. / Dry woods, fields, meadows / Leaf 5-divided

WHITE AVENS, *Geum canadense* / May–June / 16″–36″ / Dry to medium woods, disturbed ground / Leaves vary, usually 3–5 divided

AVENS, *G. aleppicum* / May–July / 20″–40″ / Wet meadows, wet woods and forest edges / Bur, center, is characteristic of yellow and white *Geums*

PRAIRIE SMOKE, *G. triflorum* / April–June / 4″–16″ / Dry to wet prairies, sandy open woods / Fruit cluster plumed, pink

PURPLE AVENS, *G. rivale* / May–July / 1′–2′ / Wet meadows and forest edges

DEWDROPS, *Dalibarda repens* / June–Sept. / Creeping, to 4″ high / Wet forests

AGRIMONY, *Agrimonia gryposepala* / July–Sept. / 1′–5′ / Medium to dry open woods and forests / Lower leaves feather-divided; fruit a stick-tight bur
(AMERICAN BURNET, *Sanguisorba canadensis*: Leaves similar; flowers in a crowded spike—white, tiny, 4-petaled; bogs east)

WOOD-SORREL FAMILY

WOOD-SORREL, *Oxalis acetosella* / June–Aug. / 3″–7″ / Wet to medium forests

YELLOW WOOD-SORREL, *O. stricta* / June–Oct. / Erect to trailing, 5″–20″ / Dry prairies, disturbed ground, sun or shade / Pods on bent stalks
(*O. europaea*: pods on straight stalks)

VIOLET WOOD-SORREL, *O. violacea* / April–June / 4″–6″ / Dry open woods, dry to medium prairies

WILD GERANIUM fl. 1″

HERB ROBERT fl. ¼″

VIRGINIA SAXIFRAGE fl. ½″

GOLDEN SAXIFRAGE fl. ⅛″

**SWAMP SAXIFRAGE
fl. ¼″**

SULLIVANTIA fl. arr. 5″

**FOAM
FLOWER
fl. arr. 3¼″**

**ALUM ROOT
fl. arr. 7″**

MITERWORT   fl. ⅛″

NAKED MITERWORT   fl. ⅛″

GERANIUM FAMILY

WILD GERANIUM, *Geranium maculatum* / April– June / 12″–28″ / Dry to medium woods

HERB ROBERT, G. *robertianum* / May–Oct. / 6″–24″ / Damp rocky shores and woods

SAXIFRAGE FAMILY

VIRGINIA SAXIFRAGE, *Saxifraga virginiensis* / April–May / 4″–16″ / Rocks, moist to dry open woods and forests (Several northern Saxifrages of exposed rocks and gravels have small fleshy leaves in rosettes or mats, resembling Sedums.)

SWAMP SAXIFRAGE, S. *pensylvanica* / May–June / 1′–3′ / Wet meadows and woods, bogs

GOLDEN SAXIFRAGE, *Chrysosplenium americanum* / April–June / 1″–3″, mats to 1′ diam. / Shady wet places, springs / Flowers green

SULLIVANTIA, *Sullivantia renifolia* / June–July / 4″–14″ / Wet shaded cliffs

ALUM ROOT, *Heuchera richardsonii* / June–July / 1′–3′ / Dry woods, dry to wet prairies (Flowering stems cut short for picture)

FOAM FLOWER, *Tiarella cordifolia* / April–May / 4″–14″ / Medium woods and forests

MITERWORT, *Mitella diphylla* / May–June / 4″–16″ / Medium to wet woods and forests, cliffs, streambanks

NAKED MITERWORT, M. *nuda* / May–June / 6″–12″ / Wet mossy forests, bogs / Stem leafless; basal leaves small, round

GRASS OF PARNASSUS, *Parnassia glauca* / Aug.–Oct. / 8″–16″ / Bogs, wet limy meadows, shores / Petals green-striped

GRASS OF PARNASSUS   fl. ½″

MARSH MALLOW fl. 5"

MUSK MALLOW fl. 2"

**FLOWER OF AN HOUR
fl. 1¾"**

POPPY MALLOW fl. 1¾"

GLADE MALLOW fl. arr. 12"

CHEESES fl. ¼"

VELVET LEAF fl. ¾″

SUNDEW fl. ⅛″

PITCHER PLANT fl. 2″

MALLOW FAMILY

MARSH MALLOW, *Hibiscus militaris* / Aug.–Sept. / 3′–7′ / Shallow water, marshes / Leaf 3-lobed
(In eastern marshes *H. palustris:* flower usually pink; stem and leaves finely hairy)

FLOWER OF AN HOUR, *H. trionum* / July–Sept. / 1′–2′ / Disturbed ground / Sepals form translucent veiny husk; flowers close by noon

MUSK MALLOW, *Malva moschata* / June–Oct. / 1′–3′ / Disturbed ground / White, lavender

CHEESES, *M. neglecta* / June–Oct. / 2″–18″, spreading to 3′ diam. / Disturbed ground, farmyards
(*Sida spinosa:* flowers yellow; leaves oblong)

GLADE MALLOW, *Napaea dioica* / June–Aug. / 3′–6′ / Wet woods and meadows, streambanks

POPPY MALLOW, *Callirhoe triangulata* / June–Aug. / 3″–24″, spreading to 4′ diam. / Dry prairies and woods, inland sands

VELVET LEAF, *Abutilon theophrasti* / July–Oct. / 1′–4′ / Disturbed ground

SUNDEW FAMILY

SUNDEW, *Drosera intermedia* / July –Aug. / 2″–8″ / Open-water bogs, shallow water, wet sunny sand / Whole plant often reddish; sticky droplets trap insects; flowers bloom one at a time
(Other Sundews: round to elongated leaves)

PITCHER PLANT FAMILY

PITCHER PLANT, *Sarracenia purpurea* / May–Aug. / 1′–2′ / Bogs / Petals fall, but petal-like sepals persist until frost; insects are trapped in the liquid at base of cup-like leaves
(Other insect-eating plants: Sundews, Bladderworts, Butterwort)

YELLOW WILD INDIGO fl. ½″

RATTLEBOX
fl. ⅜″

WILD INDIGO
fl. 1″

**PRAIRIE
WILD INDIGO**
fl. 1″

LUPINE fl. 1″

PRAIRIE TURNIP fl. ¾″

VETCH   fl. ½"

HAIRY VETCH   fl. ¾"

PURPLE VETCH   fl. ¾"

BEAN FAMILY

WILD INDIGO, *Baptisia leucantha* / June–July / 3'–6' / Dry to medium prairies and open woods, sandy soil / Erect bushy plant

PRAIRIE WILD INDIGO, *B. leucophaea* / May–June / 10"–32" / Dry prairies and open woods, sandy soil / Low drooping bushy plant

YELLOW WILD INDIGO, *B. tinctoria* / June–July / 1'–3' / Dry or sandy soil—east / Baptisias have inflated pods

RATTLEBOX, *Crotalaria sagittalis* / June–Sept. / 4"–16" / Dry disturbed soil / Leaves not divided; stem winged; pods inflated

LUPINE, *Lupinus perennis* / May–June / 8"–24" / Dry sandy woods and prairies

PRAIRIE TURNIP, *Psoralea esculenta* / May–July / 4"–16" / Dry sandy woods and prairies
(The bushy-branched prairie SCURF-PEAS: *P. tenuiflora,* and silky-hairy *P. argophylla* have ¼" flowers)

VETCH, *Vicia caroliniana* / May–June / to 3' / Wet to dry open woods and openings / Flower blue with white

HAIRY VETCH, *V. villosa* / June–Aug. / to 3' / Disturbed ground, sandy fields
(*V. cracca:* leaves not conspicuously hairy; *V. angustifolia:* flowers in 2's near leaf bases)

PURPLE VETCH, *V. americana* / June–Aug. / to 3' / Moist open woods and meadows / 8-18 leaflets

(Note: Vetches, Peas, and Vetchlings have leaf-tip tendrils, with which they climb or trail. Vetches have crowded leaflets.)

BEACH PEA   fl. 1"

WILD PEA   fl. ¾"

VETCHLING   fl. ¾"

TICK-TREFOIL   fl. ⅜"

PALE VETCHLING   fl. ¾"

GOAT'S RUE   fl. ⅝"

MILK-VETCH   fl. ½″

YELLOW SWEET-CLOVER
                                fl. ¼″

CANADA TICK-CLOVER   fl. ¼″

BEAN FAMILY

BEACH PEA, *Lathyrus maritimus* / June–Aug. / Trailing to 3′ / Shores and beaches of the Great Lakes, inland sands / Leaves slightly fleshy

WILD PEA, *L. palustris* / June–July / 1′ 4′ / Wet meadows, shores, marshes / 4–8 leaflets, wide to narrow
(Leaflets 2: *L. tuberosus:* fragrant; *L. latifolius:* stem winged; *L. pratensis:* flower yellow—all 3 of disturbed ground)

VETCHLING, *L. venosus* / June–July / 2′–3′ / Wet to medium woods, prairies, streambanks

PALE VETCHLING, *L. ochroleucus* / May–July / 2′–3′ / Wet to dry woods and forests, cliffs

GOAT'S RUE, *Tephrosia virginiana* / June–July / 8″–28″ / Dry open woods, inland sands / Plant usually gray-hairy; leaflets long-oval

TICK-TREFOIL, *Desmodium glutinosum* / July–Aug. / 2′–5′ / Dry to medium woods / Leaves form bushy mass; leaflets broad with abrupt slender tips
(*D. nudiflorum:* flowering stem leafless, leaflets not slender-tipped; *D. pauciflorum:* flowers white, often among leaves; *D. rotundifolium:* trailing, leaflets round—southeast)

CANADA TICK-CLOVER, *D. canadense* / July–Aug. / 3′–6′ / Wet to medium prairies, streambanks / Flowers close together on short, numerous racemes
(Dry prairies—*D. illinoense:* flowers far apart on a few long racemes, stem very sticky-hairy; woodland and wood-edges—*D. cuspidatum* and others: stems leafy, pod-sections triangular)

Note: Desmodiums have flat pods, breaking into round or angular stick-tight segments.

MILK-VETCH, *Astragalus canadensis* / July–Aug. / 1′–4′ / Wet to medium open woods, prairies, shores
(*A. neglectus:* flowers not crowded, on limy cliffs and gravel shores; *A. caryocarpus:* flowers purple, prairies—west)

YELLOW SWEET-CLOVER, *Melilotus officinalis* / May–Sept. / 1′–6′ / Disturbed ground
(The similar *M. alba:* white flowers)

BEAN FAMILY

RABBITFOOT CLOVER, *Trifolium arvense* / May–Sept. / 4″–16″ / Dry disturbed ground / Flower head hairy, lavender-gray

RED CLOVER, *T. pratense* / May–Sept. / 1′–3′ / Disturbed ground / Flowers red, pink

ALSIKE, *T. hybridum* / May.–Sept. / 1′–3′, erect or sprawling / Disturbed ground / White flowers turn pink (WHITE DUTCH CLOVER, *T. repens:* stems creeping, rooting; flowers turn pinkish or brown; common in lawns)

HOP CLOVER, *T. agrarium* / May–Sept. / 8″–20″ / Disturbed ground

BLACK MEDIC, *Medicago lupulina* / May–Sept. / 1″–12″, spreading to 3′ diam. / Disturbed ground / Pods curved, black

ALFALFA, *M. sativa* / June–Sept. / 1′–3′ / Disturbed ground / Purple, blue, yellow, green, white; pods coiled

WHITE PRAIRIE-CLOVER, *Petalostemum candidum* / June–July / 1′–3′ / Medium to dry prairies, dry open woods

PURPLE PRAIRIE-CLOVER, *P. purpureum* / June–Aug. / 1′–3′ / Medium to dry prairies, dry open woods (SILKY PRAIRIE-CLOVER, *P. villosum:* whole plant gray-woolly, dry prairies)

LEAD PLANT, *Amorpha canescens* / June–July / 8″–40″ / Medium to dry open woods and prairies / Plant usually gray-hairy; leaflets short-oval (INDIGO-BUSH, *A. fruticosa*: 3′–12′, riverbanks)

BIRDFOOT TREFOIL, *Lotus corniculatus* / June–Aug. / 6″–24″ / Disturbed ground

AXSEED, *Coronilla varia* / May–Sept. / 1′–2′ / Disturbed ground / Plant bushy

RABBITFOOT CLOVER fl. arr. 1″

WHITE PRAIRIE-CLOVER fl. arr. 2″

PURPLE PRAIRIE-CLOVER fl. arr. 1½″

 LEAD PLANT fl. arr. 3″

RED CLOVER fl. arr. 1⅛″

**BLACK MEDIC
fl. arr. ¼″**

ALSIKE fl. ¼″

ALFALFA fl. ½″

**HOP CLOVER
fl. ¼″**

**BIRDFOOT TREFOIL
fl. ½″**

AXSEED fl. ½″

ROUNDHEAD BUSH-CLOVER   fl. ⅜″

CREEPING BUSH-CLOVER   fl. ⅜″

HOG PEANUT   fl. ½″

BUSH-CLOVER
fl. ⅜″

GROUND NUT
fl. ½″

WILD SENNA
fl. ¾″

PARTRIDGE
PEA   fl. ½″

BEAN FAMILY

BUSH-CLOVER, *Lespedeza virginica* / July–Sept. / 1'–4' / Sandy or rocky open woods / Leaflets slender

ROUNDHEAD BUSH-CLOVER, *L. capitata* / July–Sept. / 2'–4' / Dry to medium open woods and prairies, sandy soil / Old stems persistent
(*L. hirta:* flower-heads stalked, plant slender)

CREEPING BUSH-CLOVER, *L. repens* / 1"–6", matting to 2' diam. / Sandy or rocky open woods
(In similar habitat and stalked groups of purple flowers: VIOLET BUSH-CLOVER, *L. violacea*—bushy, 1'–4')

HOG PEANUT, *Amphicarpa bracteata* / Aug.–Sept. / Twining 6"–60" / Wet to dry woods and forests, meadows, prairies / White, lavender; plant varies in hairiness
(WILD BEAN, *Strophostyles helvola:* leaflets lobed; *S. leiosperma:* plant gray-hairy—both twining)

RED MILKWORT fl. arr. ¾"

GROUND NUT, *Apios americana* / July–Aug. / Twining to 7' / Wet woods, streambanks

SENNA FAMILY

WILD SENNA, *Cassia hebecarpa* / July–Aug. / 3'–6' / Medium open woods, streambanks

MILKWORT fl. arr. ½"

PARTRIDGE PEA, *C. fasciculata* / July–Sept. / 4"–36" / Dry meadows and prairies, inland sands
(*C. nictitans:* flowers ¼"—south and east; *Desmanthus illinoensis:* flowers crowded, tiny, whitish; pods in round heads. All 3 have 'sensitive plant' leaves, but close only at night.)

MILKWORT FAMILY

RED MILKWORT, *Polygala sanguinea* / July–Sept. / 4"–16" / Open woods and prairies, moist meadows
(*P. incarnata:* leaves cylindric, falling off; *P. cruciata:* leaves whorled)

MILKWORT, *P. verticillata* / July–Oct. / 2"–16" / Medium prairies and woods / Leaves whorled to alternate; white, green, or purple

SENECA SNAKEROOT, *P. senega* / May–July / 4"–20" / Dry to wet woods and prairies / Grows in clumps; leaves narrow to broad-oval

SENECA SNAKEROOT
fl. arr. 1½"

GAYWINGS fl. ¾"
SAND MILKWORT fl. arr. 3"

WHITE VIOLET fl. ¼"

SMALL WHITE VIOLET
fl. ¼"

EARLY YELLOW VIOLET
fl. ⅜"

WOOLLY BLUE VIOLET fl. ½"

ARROWLEAF VIOLET fl. ¼"

LANCELEAF VIOLET
fl. ¼"

BIRDFOOT VIOLET fl. ⅝″

PRAIRIE VIOLET fl. ½″

BUTTERFLY VIOLET fl. ½″

BLUE MARSH VIOLET fl. ½″

MILKWORT FAMILY

GAYWINGS, Polygala paucifolia / May–June / 3″–6″ / Wet to medium forests / Grows in patches; leaves evergreen, often pink to purple beneath

SAND MILKWORT, P. polygama / June–Aug. / 4″–10″ / Dry sandy soil, dry prairies / Grows in clumps

VIOLET FAMILY, blooming from April to June, occ. again in Sept.

EARLY YELLOW VIOLET, Viola rotundifolia / Moist woods—east / Leaves rounded, often lie flat on ground

WHITE VIOLET, V. incognita / Dry to wet woods and forests / Flowers fragrant, soon hidden by leaves; may form patches

SMALL WHITE VIOLET, V. pallens / Wet forests, bogs, streambanks / Flowers fragrant; leaves hairless

WOOLLY BLUE VIOLET, V. sororia / Wet to medium woods and meadows / Flowers often hidden by the *hairy* leaves

LANCELEAF VIOLET, V. lanceolata / Bogs, moist sandy meadows / Leaves very narrow

ARROWLEAF VIOLET, V. sagittata / Wet to dry meadows, prairies, inland sands / Flowers purple to white; plant hairless to finely-hairy

BLUE MARSH VIOLET, V. cucullata / Wet meadows and woods / Flowers light to dark blue, often held above leaves; leaves often curled; plant hairless

BUTTERFLY VIOLET, V. papilionacea / Wet to medium woods, wet meadows / Flowers barely above leaves; many color variations: white with dark blue center, magenta, speckled

BIRDFOOT VIOLET, V. pedata / Dry open woods and prairies, inland sands / Flowers, appearing *flat,* often hide leaves—blue, white, or bi-colored; leaves deeply divided

PRAIRIE VIOLET, V. petatifida / Wet to dry prairies / Flowers few, *not* flat, above deeply divided leaves

CANADA VIOLET fl. ⅜″

STRIPED VIOLET fl. ⅜″

YELLOW VIOLET fl. ⅜″

SAND VIOLET fl. ¼″

LONGSPUR VIOLET fl. ¼″

GREEN VIOLET fl. ¼″

VIOLET FAMILY

The Violets on the opposite page all bloom **FROM LEAFY STEMS**, not from the ground

CANADA VIOLET, *Viola canadensis* / Medium woods and forests / Flowers pink-backed
(*V. rugulosa:* leaves velvety beneath)

STRIPED VIOLET, *V. striata* / Wet to medium woods and meadows, streambanks / Leaves small, dark green

SAND VIOLET, *V. adunca* / Moist to dry sand or gravel soil, moist open woods / Spur ¼", conspicuous

LONGSPUR VIOLET, *V. rostrata* / Medium woods and forests / Plant rarely over 5" high; spur ½", conspicuous
(GREAT-SPUR VIOLET, *V. selkirkii:* flower not from leafy stem; forests)

YELLOW VIOLET, *V. pubescens* / Dry to medium woods and forests / Stems and leaves vary from smooth to hairy

GREEN VIOLET, *Cubelium concolor* / May—June / 12"–40" / Moist woods, ravines

JEWELWEED FAMILY

SPOTTED JEWELWEED, *Impatiens biflora* / July–Sept. / 2′–6′ / Wet to medium woods and forests, wet meadows, streambanks / Orange to yellow, occ. white or pink
(*I. pallida:* pale yellow, sac broader than long)

PARSLEY FAMILY

SWEET CICELY, *Osmorhiza claytoni* / May–June / 1′–3′ / Dry to wet woods, wet to medium forests / Anise scented; seeds slender, black, sharp-spined

HONEWORT, *Cryptotaenia canadensis* / June–July / 1′–3′ / Wet to dry woods / Flowers on stalks of unequal length; leaves variable in shape

SPOTTED JEWELWEED   fl. 1″

SWEET CICELY   fl. arr. 3″

HONEWORT   fl. arr. 2″

PARSLEY FAMILY

AMERICAN CHERVIL, *Chaerophyllum procumbens* / May / 6"–20" / Wet to moist woods

BLACK SNAKEROOT, *Sanicula gregaria* / June–July / 1'–3' / Dry to medium woods / Fruit a small bur; basal leaves 7–9 divided

HARBINGER OF SPRING, *Erigenia bulbosa* / Mar.–May / 2"–10" / Medium woods

RATTLESNAKE MASTER, *Eryngium yuccifolium* / July–Aug. 1½'–3' / Dry to medium prairies, open woods / Leaves spiny-edged, yucca-like

COWBANE, *Oxypolis rigidior* / July–Sept. / 2'–4' / Wet meadows and prairies, wet woods / Leaflets long, slender, usually not toothed

WATER-HEMLOCK, *Cicuta bulbifera* / July–Sept. / Marshes, streambanks / Bulblets take place of most flowers

SPOTTED WATER-HEMLOCK, C. *maculata* / June–Aug. / 2'–7' / Wet meadows and woods edges, marshes (Hemlocks, like cowbanes, poisonous to eat, include 2 others with *untoothed* feather-divided leaves: *Conioselinum chinense and Conium maculatum*)

WATER-PARSNIP, *Sium suave* / July–Aug. / 2'–6' / Shallow water, shores / Above-water leaves sharp-toothed, submerged finely-divided

SMALL WATER-PARSNIP, *Berula erecta* / Aug.–Sept. / 8"–36" / Shallow water of springs, cold streams / Leaflets divided or lobed

QUEEN ANNE'S LACE, *Daucus carota* / June–Sept. / 1'–5' / Disturbed ground / Central flower purple; basal leaves finely divided; carrot odor
(CARAWAY, *Carum carvi:* May–July; leaves sparsely divided; disturbed ground)

ANGELICA, *Angelica atropurpurea* / June–Aug. / 2'–8' / Wet meadows, woods and forest edges, streambanks / Stems often purple

COW PARSNIP, *Heracleum lanatum* / June–July / 3'–7' / Moist disturbed ground, streambanks / Leaves white-woolly beneath

AMERICAN CHERVIL fl. arr. 1"

BLACK SNAKEROOT fl. arr. ¾"

HARBINGER OF SPRING fl. arr. ¾"

RATTLESNAKE MASTER fl. arr. 1"

COWBANE
fl. arr. 3″

WATER-HEMLOCK
fl. arr. 2″

SPOTTED WATER-HEMLOCK
fl. arr. 3″

SMALL WATER-PARSNIP
fl. arr. 2″

WATER-PARSNIP   fl. arr. 3″

QUEEN ANNE'S LACE
fl. arr. 4½″

ANGELICA
fl. arr. 6″

COW PARSNIP   fl. arr. 6″

YELLOW PIMPERNEL fl. arr. 3″

PARSNIP fl. arr. 5″

**HEARTLEAF
GOLDEN ALEXANDERS
fl. arr. 2″**

PARSLEY FAMILY

YELLOW PIMPERNEL, *Taenidia integerrima* / May–June / 16″–32″ / Dry woods, prairies /Leaves without hairs, not toothed
(PRAIRIE PARSLEY, *Polytaenia nuttallii:* umbels dense, domed; leaflets wedge-shaped, cut in at margin)

PARSNIP, *Pastinaca sativa* / June–July / 1′–5′ / Disturbed ground, meadows / Stem angular; juice can blister skin

HEARTLEAF GOLDEN ALEXANDERS, *Zizia aptera* / May–June/ 1′–3′ / Wet meadows, wet to dry prairies, open woods, shores / Basal leaves not divided
(*Z. aurea:* all leaves divided, coarsely-toothed, stalked; MEADOW PARSNIP, *Thaspium trifoliatum:* leaves finely-toothed, almost stalkless—flowers yellow or purple)

GINSENG FAMILY

WILD SARSAPARILLA, *Aralia nudicaulis* / May–June / 8″–20″ / Dry to wet woods and forests

BRISTLY SARSAPARILLA, *A. hispida* / June–July / 6″–36″ / Dry forests, cliffs, inland sands / Stem prickly

SPIKENARD, *A. racemosa* / July / 3′–7′ / Dry to medium woods, medium forests / Leaves large, coarse

GINSENG, *Panax quinquefolium* / July / 8″–24″ / Dry to medium woods, medium forests / Flowers green, in rounded cluster

DWARF GINSENG, *P. trifolium* / May–June / 4″–8″ / Wet to medium forests

BUCKTHORN FAMILY

NEW JERSEY TEA, *Ceanothus americanus* / June–Aug. / 1′–3′ / Open woods, medium prairies / Low woody shrub; flowers among leaves
(*C. ovatus:* flowers terminal, leaf narrow-oval, dry prairies)

SANDALWOOD FAMILY

STAR TOADFLAX, *Comandra umbellata* / May–July / 4″–12″ / Dry open woods and prairies, shores

BRISTLY SARSAPARILLA fl. arr. 1″

WILD SARSAPARILLA fl. arr. 1¼″

GINSENG berry ¼″

DWARF GINSENG fl. arr. ¾″

SPIKENARD fl. arr. 5″

NEW JERSEY TEA fl. arr. 3″

STAR TOADFLAX fl. ¼″

**SHEEP SORREL
fl. arr. 5″**

PALE DOCK   fl. ¼″

MARSH PEPPER   fl. arr. 3″

CURLY DOCK   fl. arr. 10″

SMARTWEED   fl. arr. 4″

**WATER SMARTWEED
fl. arr. 1½″**

JOINTWEED   fl. ⅛″

KNOTWEED   fl. 1/16″

PINKWEED   fl. arr. 2″

SMARTWEED FAMILY

CURLY DOCK, *Rumex crispus* / June–Aug. / 1′–4′ / Disturbed ground / Leaf crinkly-edged (BLUNTLEAF DOCK, *R. obtusifolius:* leaves broad, woods and forests)

PALE DOCK, *R. altissimus* / July–Aug. / 1′–3½′ / Disturbed ground / In patches (Other Docks may be found in wetlands)

SHEEP SORREL, *R. acetosella* / May–Aug. / 4″–12″ / Disturbed ground, wet to dry

MARSH PEPPER, *Polygonum hydropiper* / July–Sept. / 6″–24″ / Marshes, streambanks, shores / Flowers green, minutely dotted (WHITE SMARTWEED, *P. punctatum:* flowers dotted, but white)

WATER SMARTWEED, *P. natans* / June–Aug. / 1′–5′ / Plant erect, in wet meadows or shallow water, or floating; flower arrangement blunt

SMARTWEED, *P. coccineum* / June–Sept. / 1′–5′ / Plant erect, in wet meadows or shallow water, or floating; flower arrangement pointed

PINKWEED, *P. pensylvanicum* / July–Sept. / 2′–6′ / Disturbed ground, often wet / Flowers pink to white, on minutely sticky-hairy stem (2 similar common species *without* sticky hairs: LADY'S THUMB, *P. persicaria* and DOCKLEAF SMARTWEED, *P. lapathifolium,* with drooping flower spikes)

KNOTWEED, *P. aviculare* / Aug.–Oct. / 1″–6″, often forming mats to 18″ diam / Disturbed ground

JOINTWEED, *Polygonella articulata* / July–Oct. / 4″–16″ / Dry sandy soil, cliffs / Flowers stalked (SLIM KNOTWEED, *Polygonum tenue:* plant very slender, with sparse tiny white flowers not stalked)

BLACK BINDWEED   fl. arr. 3½"

SMARTWEED FAMILY

BLACK BINDWEED, *Polygonum scandens* / July–Oct. / Twining to 5' / Wet woods and forests, meadows, dry cliffs
(FALSE BUCKWHEAT, *P. cilinode:* erect or trailing; stem often reddish; disturbed ground of forest clearings, shores)

ARROWLEAF TEAR THUMB, *P. sagittatum* / Aug.–Oct. / Climbing or tangling to 6' / Wet meadows, marshes, streambanks / Stem sharp-spiny
(*P. arifolium:* leaf lobes spread)

JUMP SEED, *P. virginianum* / July–Sept. / 20"–40" / Medium woods, cliffs

ARROWLEAF
TEAR THUMB
fl. ⅜"

NETTLE FAMILY

PELLITORY, *Parietaria pensylvanica* / May–Sept. / 4"–16" / Dry woods, shady disturbed ground

STINGING NETTLE, *Urtica dioica* / June–Sept. / 1½'–6' / Dry to wet disturbed ground / Plant has stinging hairs; stem square; leaves opposite

CLEARWEED, *Pilea pumila* / July–Sept. / 4"–20" / Medium to wet woods / Stem translucent; leaves have 3 principal veins

FALSE NETTLE, *Boehmeria cylindrica* / Aug.–Sept. / 16"–40" / Wet woods, shores, marshes

WOOD NETTLE, *Laportea canadensis* / July–Aug. / 16"–40" / Wet to medium woods and forests / Plant has stinging hairs; leaves alternate

LIZARD TAIL FAMILY

LIZARD TAIL, *Saururus cernuus* / June–Aug. / 2'–4' / Shallow water, marshes, wet woods

JUMPSEED   fl. ¼"

PELLITORY fl. 1/8"

STINGING NETTLE
fl. 1/8"

CLEARWEED fl. 1/8"

WOOD NETTLE fl. 1/8"

FALSE NETTLE fl. 1/8"

LIZARD TAIL fl. arr. 6"

FLOWERING SPURGE
fl. ¼″

3-SEEDED MERCURY   fl. 1/16″

SPOTTED SPURGE   fl. 1/16″

LEAFY SPURGE   fl. arr. 5″

POKEWEED   fl. arr. 6″

FALSE MERMAID   fl. ⅛″

66—67

SPURGE FAMILY

FLOWERING SPURGE, *Euphorbia corollata* / June–Sept. / 12"–40" / Dry open woods and prairies, inland sands, disturbed ground

LEAFY SPURGE, *E. esula* / May–Sept. / 12"–28" / Fields, roadsides, disturbed ground / Forms patches (Narrow-leaved CYPRESS SPURGE, *E. cyparissias:* 8"–16", common garden escape)

SPOTTED SPURGE, *E. maculata* / May–Sept. / 1"–3", forming mats to 18" diam. / Disturbed ground / Leaves often have maroon spot

3-SEEDED MERCURY, *Acalypha rhomboidea* / July–Oct. / 8"–24" / Wet to medium woods, disturbed ground / Juice not milky; flowers surrounded by lobed cup-like leaves

POKEWEED FAMILY

POKEWEED, *Phytolacca americana* / July–Sept. / 2'–9' / Medium woods, disturbed ground / Plant bushy

FALSE MERMAID FAMILY

FALSE MERMAID, *Floerkea proserpinacoides* / May–June / 1"–6" / Medium woods / Plant weak, tangling

GOOSEFOOT FAMILY

WHITE GOOSEFOOT, *Chenopodium album* / June–Oct. / 6"–40" / Disturbed ground / Leaves at first white-mealy, coarse-toothed

GOOSEFOOT, *C. hybridum* / June–Oct. / 6"–40" / Disturbed ground in medium to dry woods and forests / Leaves with few coarse teeth

SUMMER CYPRESS, *Kochia scoparia* / July–Sept. / 6"–40" / Disturbed ground / Plant bushy, may turn red in fall

GOOSEFOOT   ball ⅛"

SUMMER CYPRESS   fl. 1/16"

STRAWBERRY BLITE   ball ⅝″

WINGED PIGWEED   plant 14″

CREEPING AMARANTH   fl. 1/16″

RUSSIAN THISTLE   stem ¼″

SPEARSCALE   fl. 1/16″

REDROOT PIGWEED
fl. arr. 2″

COTTONWEED fl. arr. 2″

GOOSEFOOT FAMILY

STRAWBERRY BLITE, *Chenopodium capitatum* / July–Sept. / 8″–24″ / Sunny disturbed ground in forests / Stem weak, drooping

CARPETWEED   fl. ⅛″

WINGED PIGWEED, *Cycloloma atriplicifolia* / July–Aug. / 4″–32″ / Dry sandy soil / Plant turns purple in fall

RUSSIAN THISTLE, *Salsola kali* / July–Oct. / 1′–2½′ / Dry disturbed ground / Plant bushy; stem pink-striped; leaves needle-like

SPEARSCALE, *Atriplex patula* / July–Oct. / 6″–40″, erect or sprawling / Disturbed ground / Leaves opposite; flower sandwiched in a pair of small flat leaves

AMARANTH FAMILY

CREEPING AMARANTH, *Amaranthus graecizans* / July–Sept. / 2″–24″ / Disturbed ground / Plant sprawls

REDROOT PIGWEED, *A. retroflexus* / July–Sept. / 8″–60″ / Disturbed ground / Stem often red

COTTONWEED, *Froelichia floridana* / July–Sept. / 2′–5′ / Dry sunny sandy soil / Leaves oposite

CARPETWEED FAMILY

CARPETWEED, *Mollugo verticillata* / June–Sept. / 1″, forming mats to 12″ diam. / Dry disturbed ground

HOPS   fl. arr. 5½″

HEMP   fl. arr. 6″

MULBERRY FAMILY

HOPS, *Humulus lupulus* / July–Aug. / Twining vine to 30′ / Disturbed ground, woods, fencerows / Flowers on male plant (not pictured) tiny, diffuse

HEMP, *Cannabis sativa* / June–Oct. / 8″–40″ / Disturbed ground, sun or shade / Upper plant parts have strong odor

POISON IVY   fl. arr. 4″

POISON IVY   leaves

SUMAC FAMILY
POISON IVY, *Rhus radicans* / May–June / 4″–36″, or a tall woody climber particularly in wet or damp forests and woods / In almost all habitats / Leaflets THREE, toothed or lobed, dull or shiny, large or small, turning yellow or red; small flowers often hidden by leaves; berries dry, yellowish-white; lower stem brown, woody

BLUEBELL FAMILY
TALL BELLFLOWER, *Campanula americana* / July–Sept. / 3′–6′ / Medium to wet woods, streambanks (BELLFLOWER, *C. rapunculoides:* flowers hang bell-like in long crowded racemes; basal leaves violet-shaped; garden escape)

MARSH BELLFLOWER, *C. aparinoides* / June–Aug. / 6″–36″, tangling / Wet meadows, streambanks / White to blue; stems rough-clinging, like some Bedstraws

HAREBELL, *C. rotundifolia* / June–Oct. / 4″–20″ / Sandy soil, dry woods, meadows, cliffs, beaches / Early basal leaves round

VENUS' LOOKING GLASS, *Triodanis perfoliata* / May–Aug. / 4″–40″ / Dry disturbed sandy soil, cliffs / Lower flowers never open

BEDSTRAW FAMILY
BLUETS, *Houstonia caerula* / May–June / 2″–8″ / Moist meadows and prairies / Blue to white, with yellow eye

LONGLEAF BLUETS, *H. longifolia* / June–Aug. / 3″–10″ / Dry sandy or rocky soil, sun to partial shade / Slender leaves all along stem

BUTTONWEED, *Diodia teres* / July–Sept. / Erect to 4″, or sprawling to 3′ diam. / Dry sandy soil—south

PARTRIDGE BERRY, *Mitchella repens* / June–July / 1″–2″, evergreen mats to 6′ diam. / Dry to moist forests and woods / Petals 3 to 6, velvety

FOUR O'CLOCK FAMILY
WILD FOUR O'CLOCK, *Oxybaphus nyctagineus* / May–Aug. / 1′–3′ / Dry prairies, dry disturbed ground (The similar *O. hirsutus:* stems hairy)

WILD FOUR O'CLOCK   fl. ⅜″

TALL BELLFLOWER fl. ⅝"

VENUS' LOOKING GLASS fl. ⅜"

MARSH BELLFLOWER fl. ¼"

HAREBELL fl. ¾"

BLUETS fl. ⅜"

LONGLEAF BLUETS fl. ¼"

BUTTONWEED fl. ⅛"

PARTRIDGE BERRY fl. ⅜"

YELLOW BEDSTRAW
fl. arr. 1¼"

SWEET-SCENTED
BEDSTRAW fl. ⅛"

CATCHWEED fl. ⅛"

BEDSTRAW fl. ⅛"

72—73

BEDSTRAW FAMILY

YELLOW BEDSTRAW, *Galium verum* / June–Sept. / 1'–2', erect / Dry disturbed ground

CATCHWEED, G. aparine / May–June / 4"–40", climbing or sprawling / Dry to moist woods and forests / Stem and leaves harshly rough-clinging
(ROUGH BEDSTRAW, G. asprellum: leaves almost oval, very rough; meadows, swamps; July–Sept.)

BEDSTRAW, G. concinnum / June–Aug. / 5"–24", erect or tangling / Dry to medium woods / Leaves dark green, small—to ¾"

SWEET-SCENTED BEDSTRAW, G. triflorum / June–Aug. / 3"–32", sprawling / Dry to medium woods and forests / Flowers in 3's

NORTHERN BEDSTRAW, G. boreale / June–July / 6"–40", erect; often in large patches / Dry open woods and forests, prairies, meadows / Leaves in 4's rather than 6's or 8's
(Others with leaves in 4's: in wetlands: G. obtusum; G. labradoricum, leaves bent down, slender; G. tinctorium, petals only 3. Broad-leaved species, flowers green or purple: G. lanceolatum and hairy G. circaezans)

NORTHERN BEDSTRAW
fl. ⅛"

TWINFLOWER fl. ½" long

TINKER'S WEED fl. ½"

HONEYSUCKLE FAMILY

BUSH HONEYSUCKLE, *Diervilla lonicera* / June–July / ½'–3', erect to arching / Dry or rocky woods and forests, cliffs / Flowers yellow to red; leaves finely-toothed
(Other woody Honeysuckles, *Lonicera,* differ in having untoothed leaves.)

TINKER'S WEED, *Triosteum perfoliatum* / May–June / 2'–4' / Dry to medium woods / Leaves vary from slender-based to joining around stem; flowers red to greenish-yellow

TWINFLOWER, *Linnaea borealis* / June–Aug. / Flower stalks to 5" / Moist to dry forests / Flowers fragrant; plant creeping, evergreen

GOURD FAMILY, climbers

MOCK CUCUMBER, *Echinocystis lobata* / July–Sept. / Streambanks, wet thickets / Plant smooth; petals 6; fruit, 2", soft-prickly

BUR CUCUMBER, *Sicyos angulatus* / June–Aug. / Streambanks, moist soil / Plant hairy; petals 5; fruit small clustered burs

BUSH HONEYSUCKLE fl. ¾" long

MOCK CUCUMBER fl. ⅜"

BUR CUCUMBER fl. ⅜"

GAURA fl. ½"

ENCHANTER'S NIGHTSHADE fl. ¼"

SMALL SUNDROPS fl. ½"

EVENING PRIMROSE fl. 1¼"

SLENDER
EVENING PRIMROSE fl. ⅝"

SEEDBOX fl. ½"

WILLOW HERB fl. ¼"

FIREWEED fl. 1"

EVENING-PRIMROSE FAMILY

ENCHANTER'S NIGHTSHADE, *Circaea quadrisulcata* / June–Aug. / 8"–24" / Medium to dry woods / Fruit a small bur
(The NORTHERN, *C. alpina:* 4"–12", pale green delicate plant; moist forests, bogs)

GAURA, *Gaura biennis* / Aug.–Sept. / 3'–6' / Dry to moist prairies, open woods, shores / Petals turn pinkish; stems red

EVENING-PRIMROSE, *Oenothera biennis* / July–Oct. / 2'–6' / Disturbed ground, prairies / Flowering section becomes very long; stem often red

SLENDER EVENING-PRIMROSE, *O. rhombipetala* / June–Sept. / 16"–40" / Dry sandy fields and prairies / Usually many-branched from base; basal leaves feather-divided

SMALL SUNDROPS, *O. perennis* / May–July / 4"–24" / Dry to wet prairies, sandy meadows, open woods
(SUNDROPS, *O. pilosella:* fl. 1¼"—east and common in gardens)

SEEDBOX, *Ludwigia alternifolia* / July–Aug. / 20"–40" / Wet woods, shores—south
(*L. palustris:* leaves opposite; petals absent; both species have square seed capsules)

WILLOW HERB, *Epilobium coloratum* / July–Oct. / 2'–3' / Wet meadows, streambanks

FIREWEED, *E. angustifolium* / June –Aug. / 2'–6' / Wet to dry forest clearings, edges / One of the first plants to appear after burnings

BLUEBERRY FAMILY

HUCKLEBERRY, *Gaylussacia baccata* / May–June / 1'–3' / Dry to medium sandy or rocky woods and forests, bogs / Young leaves sticky beneath

HUCKLEBERRY fl. ⅜"

BLUEBERRY fl. ⅜"

INDIAN PIPE fl. 1"

PINESAP fl. ⅝"

PINEDROPS fl. ¼"

LEATHERLEAF fl. ¼"

BEARBERRY fl. ¼"

COWBERRY fl. ⅜"

CRANBERRY fl. ¼"

SNOWBERRY leaf ½"

BOG ROSEMARY fl. ¼"

BOG LAUREL fl. ⅜"

LABRADOR TEA fl. ⅜"

BLUEBERRY FAMILY

BLUEBERRY, *Vaccinium angustifolium* / May–June / 2"–14" / Woods and forest clearings, bogs, sandy or rocky soil / Leaves smooth

INDIAN PIPE, *Monotropa uniflora* / June–Sept. / 4"–8" / Dry to medium forests and woods, bogs / Flower turns up as fruit ripens; no leaves

PINESAP, *M. hypopithys* / June–Aug. / 4"–12" / Dry to moist forests, acid soil / Yellow, tan, pink, red; no leaves

PINEDROPS, *Pterospora andromedea* / June–Aug. / 1'–3' / Dry, usually pine, forests / No leaves

(The remaining family members: evergreen leaves leathery, stiff)

COWBERRY, *Vaccinium vitis-idaea* / May–June / 4"–8" / Dry peaty or rocky soil, bogs—north / Leaves oval, black-dotted beneath

CRANBERRY, *V. oxycoccus* / June–July / 2"–6", creeping / Bogs, wet sand / Leaves pointed, ¼" long, edges inrolled, whitish beneath

LEATHERLEAF, *Chamaedaphne calyculata* / April–June / 1'–3' / Bogs, wet sand / Leaves minutely scaly beneath

BEARBERRY, *Arctostaphylos uvi-ursi* / May–June / 2"–6", in mats trailing to 3' diam. / Damp to dry rocky or sandy soil, bogs / Leaves rounded at tip

SNOWBERRY, *Gaultheria hispidula* / April–May / In ground-trailing mats to 16" diam. / Mossy wet forests, bogs / ⅛" flowers hidden under oval, pointed leaves

BOG ROSEMARY, *Andromeda glaucophylla* / May–June / 8"–20" / Bogs / Leaves slender, blue-green, whitened beneath, edges inrolled

BOG LAUREL, *Kalmia polifolia* / May–June / 1'–2' / Bogs / Leaves opposite—unique in family

LABRADOR TEA, *Ledum groenlandicum* / May–June / 1½'–3' / Bogs, wet forests, cliffs / Leaves brown-woolly beneath, edges inrolled

TRAILING ARBUTUS fl. ³⁄₈ ″

WINTERGREEN fl. ¼ ″

WAX FLOWER fl. ¼ ″

PIPSISSEWA fl. ½ ″

SPOTTED PIPSISSEWA fl. ¾ ″

SIDE BELLS fl. ¼ ″

WOOD NYMPH fl. ⅝ ″

BUNCHBERRY   fl. arr. 1″

WILD GINGER   fl. ¾″

TEASEL   fl. arr. 4″

BLUEBERRY FAMILY

TRAILING ARBUTUS, *Epigaea repens* / March–May / 1″–3″ / Boggy or sandy forests and openings / Flowers fragrant; plant hairy

WINTERGREEN, *Gaultheria procumbens* / July–Aug. / 4″–8″ / Dry to medium forests, inland sands / Leaves slightly toothed

PIPSISSEWA, *Chimaphila umbellata* / July–Aug. / 4″–12″ / Dry, usually sandy, forests and woods / Leaves waxy, toothed

SPOTTED PIPSISSEWA, *C. maculata* / June–Aug. / 4″–10″ / Dry sandy woods and forests / Leaves have white splotches, toothed

WAX FLOWER, *Pyrola elliptica* / June–Aug. / 6″–12″ / Dry to medium woods and forests
(Two similar SHINLEAFS of moist forests and bogs: *P. rotundifolia,* leaves very thick and waxy; *P. asarifolia,* flowers pink. Two with leaves shorter than leaf-stalks: *P. minor,* flower globe-like; *P. virens,* flower greenish)

SIDE BELLS, *P. secunda* / June–July / 4″–8″ / Wet to medium forests, bogs

WOOD NYMPH, *Moneses uniflora* / July–Aug. / 1″–5″ / Wet forests, bogs

DOGWOOD FAMILY

BUNCHBERRY, *Cornus canadensis* / May–July—occ. Oct. / 2″–8″ / Wet to moist forests, bogs

BIRTHWORT FAMILY

WILD GINGER, *Asarum canadense* / April–May / Leaves 2″–8″ wide / Medium woods and forests / Stems creeping

TEASEL FAMILY

TEASEL, *Dipsacus laciniatus* / July–Sept. / 2′–7′ / Disturbed ground / Plant sharp-prickly; flowers white to lavender; leaf-bases join around stem to form a cup
(*D. sylvestris:* leaves slender, not feather-lobed)

GENTIAN FAMILY

BITTERBLOOM, *Sabatia angularis* /
July–Aug. / 1′–3′ / Moist woods,
moist sand or peat—south

COLUMBO, *Swertia caroliniensis* /
May–June / 3′–6′ / Medium limy
woods—south / Leaves in 4's; stem
leafy

BOGBEAN, *Menyanthes trifoliata* /
April–July / 4″–12″ / Bogs, sloughs,
wet forests / Stems long, floating
or creeping; flower fuzzy

BOTTLE GENTIAN, *Gentiana an-
drewsii* / Aug.–Oct. / 1′–3′ / Wet
to medium meadows and prairies,
wet woods / Flowers dark to light
blue, rosy lavender, or white; petal
tips shorter than folds
(With petal tips *longer* than folds: RED-
STEM G., *G. rubricaulis,* smoky lavender flowers, meadows and shores—
north; CLOSED G., *G. clausa:* sepals
hairy at margin, meadows and wood-
edges—east. Hybrids between these 3
and Downy and Cream G. occur)

CREAM GENTIAN, *G. flavida* / Aug.–
Sept. / 2′–3′ / Dry to medium mead-
ows and prairies, open woods /
Petal tips longer than folds

DOWNY GENTIAN, *G. puberula* /
Aug.–Oct. / 8″–20″ / Dry sandy or
limy prairies, open woods

STIFF GENTIAN, *G. quinquefolia* /
Aug.–Oct. / 3″–16″ / Wet to dry limy
prairies and woods

FRINGED GENTIAN, *G. crinita* /
Aug.–Oct. / 12″–32″ / Wet meadows,
damp woods, streambanks / Leaves
⅓ as broad as long
(LESSER FRINGED G., *G. procera:*
leaves very slender, moist limy meadows
and rocky shores)

SPUR GENTIAN, *Halenia deflexa* /
July–Aug. / 4″ 36″ / Wet to medium
forests, bogs / 4 nectar spurs make
flower appear square

BITTERBLOOM   fl. ⅞ ″

COLUMBO   fl. ⅞ ″

BOGBEAN   fl. ¾ ″

BOTTLE GENTIAN fl. 1¼″

CREAM GENTIAN fl. 1¼″

DOWNY GENTIAN fl. 1″

FRINGED GENTIAN fl. 1⅛″

STIFF GENTIAN fl. ⅝″

SPUR GENTIAN fl. ⅜″

JACOB'S LADDER fl. ⅜″

WILD BLUE PHLOX fl. ¾″

DOWNY PHLOX fl. ¾″

TENPOINT PHLOX fl. ½″

HOARY VERVAIN fl. ¼″

**BLUE VERVAIN
fl. ¼″**

PHLOX FAMILY

JACOB'S LADDER, *Polemonium reptans* / May–June / 8″–16″ / Moist meadows, low woods edges / Basal leaves feather-divided, ladder-like

WILD BLUE PHLOX, *Phlox divaricata* / April–June / 6″–24″ / Medium to wet woods and forests, cliffs / Blue to white; basal leaves remain green in winter

DOWNY PHLOX, P. *pilosa* / April–June / 6″–24″ / Dry to wet prairies and open woods, inland sands / Plant finely-haired
(*P. glaberrima:* not hairy, June–Aug.; *P. maculata:* stem purple-spotted, wet meadows and woods; *P. paniculata:* garden perennial, and woods—south; *Collomia linearis:* flowers lilac to white, leaves alternate, dry disturbed ground —north)

TENPOINT PHLOX, P. *bifida* / April–May / 4″–12″, mat-forming / Dry sandy soil, cliffs—southwest
(MOSS PHLOX, *P. subulata:* petals scarcely notched; leaves crowded; flowers magenta, lavender, or white—east and garden escape)

VERBENA FAMILY

HOARY VERVAIN, *Verbena stricta* / July–Sept. / 8″–36″ / Dry disturbed ground, prairies, inland sands / Flower spike elongates to 8″
(*V. simplex:* leaf narrow, plant scarcely hairy)

BLUE VERVAIN, V. *hastata* / July–Oct. / 16″–40″ / Moist meadows, fields, prairies, streambanks / Flower spike elongates to 8″

VERVAIN, V. *bracteata* / May–Oct. / 1″–12″, often creeping / Dry prairies and disturbed ground

WHITE VERVAIN, V. *urticifolia* / July–Oct. / 16″–40″ / Edges and disturbed ground of woods / Flower spike elongates to 1′

FOGFRUIT, *Phyla lanceolata* / May–Oct. / 2″–32″ / Streambanks, wet forests / Flower spike elongates to 1½″

VERVAIN fl. ⅛″

WHITE VERVAIN fl. ⅛″

FOGFRUIT fl. arr. 1″

PRIMROSE FAMILY

ROCK JASMINE, *Androsace occidentalis* / April / ½"–20" / Cliffs, inland sands—west / Plant turns red

SHOOTING STAR, *Dodecatheon meadia* / April–June / 8"–24" / Medium to dry woods and prairies / White, pale pink; leaf base reddish (AMETHYST SHOOTING STAR, *D. radicatum:* flower deep pink; leaf bases not reddish; damp or shady limy cliffs and streambanks)

ROCK JASMINE   fl. ⅛"

SHOOTING STAR   fl. ⅞"

STAR FLOWER, *Trientalis borealis* / May–June / 4"–8" / Wet to medium forests, bogs / Petals 7
(Others of Primrose F. with star-like flowers: SCARLET PIMPERNEL, *Anagallis arvensis,* flowers red or blue among opposite leaves; WATER PIMPERNEL, *Samolus parviflorus:* flowers white, leaves opposite, in shallow water or mud)

FRINGED LOOSESTRIFE, *Lysimachia ciliata* / June–July / 10"–40" / Wet woods, streambanks / Leaf-stalk fringed with white hairs

LANCELEAF LOOSESTRIFE, *L. lanceolata* / June–July / 3"–24" / Medium to dry open woods and prairies —south / Patches to 6' diam.
(*L. hybrida:* 2'–5', clumps; shores)

PRAIRIE LOOSESTRIFE, *L. quadriflora* / July–Aug. / 1'–2' / Wet limy meadows, wet to medium prairies / Leaves very narrow

4-LEAF LOOSESTRIFE, *L. quadrifolia* / June–July / 1'–3' / Medium to dry open woods and forests / Leaves whorled in 3's to 6's; forms patches

TUFTED LOOSESTRIFE, *L. thyrsiflora* / May–July / 1'–2½' / Edges of bogs and marshes, shallow water

SWAMP CANDLES, *L. terrestris* / June–Aug. / 16"–32" / Wet shores, bog edges, shallow water

MONEYWORT, *L. nummularia* / June–Aug. / 1"–2", creeping and matting / Wet woods, disturbed ground

BIRDEYE PRIMROSE, *Primula mistassinica* / 1"–3½" / Gravel shores, rock cliffs—north

STAR FLOWER   fl. ⅝"

FRINGED LOOSESTRIFE fl. 5/8″

LANCELEAF LOOSESTRIFE fl. 5/8″

BIRDEYE PRIMROSE fl. 5/8″

4-LEAF LOOSESTRIFE
fl. 5/8″

PRAIRIE LOOSESTRIFE fl. 1/2″

MONEYWORT fl. 5/8″

TUFTED LOOSESTRIFE
fl. arr. 1″

SWAMP CANDLES
fl. arr. 12″

VIRGINIA BLUEBELL   fl. ½"

FORGET ME NOT   fl. ¼"

SMALL BUGLOSS   fl. ¼"

BORAGE FAMILY

VIRGINIA BLUEBELL, *Mertensia virginica* / May–June / 12"–26" / Medium to wet woods / Buds pink; flowers hang down; plant not hairy (*M. paniculata:* plant hairy, damp forests—northwest)

FORGET ME NOT, *Myosotis scorpioides* / May–Sept. / 4"–24" / Wet soil, streambanks, shallow water (Wetland, *M. laxa;* dry or disturbed soil: *M. verna,* white; *M. arvensis,* blue)

SMALL BUGLOSS, *Lycopsis arvensis* / June–Aug. / 6"–24" / Sunny dry disturbed ground / Plant spiny

STICKSEED, *Hackelia virginiana* / July–Sept. / 1'–3' / Dry to medium woods and forests / Fruit a small hanging 4-parted bur; flower white (*Lappula echinata:* flowers blue, not white; fruit an erect bur; sunny disturbed ground)

FALSE GROMWELL, *Onosmodium molle* / June–July / 12"–40" / Sand or rock prairies, dry open woods / Plant bristly

HOUND'S TONGUE, *Cynoglossum officinale* / June–July / 8"–48" / Dry open woods, disturbed ground / Plant hairy; fruit a 4-parted bur (NORTHERN COMFREY, *C. boreale:* leaves sparse, flowers blue, forests. Comfreys: *Symphytum officinale,* hairy; *S. asperum,* spiny flowers 1/2")

BLUEWEED, *Echium vulgare* / June–Sept. / 1'–3' / Disturbed ground / Plant bristly

PUCCOON, *Lithospermum caroliniense* / May–July / 6"–24" / Dry prairies and open woods, inland sands / Plant bristly

HOARY PUCCOON, *L. canescens* / April–June / 4"–18" / Dry to moist prairies, dry open woods / Flower deep orange; plant finely-hairy

GROMWELL, *L. incisum* / May–June / 4"–20" / Dry prairies, sandy or rocky soil / Petals fringed, yellow

CORN GROMWELL, *L. arvense* / / May–July / 4"–32" / Disturbed ground / Flower white (*L. latifolium:* leaves 1"–2" broad)

STICKSEED   fl. ⅛″

FALSE GROMWELL   fl. ⅝″

BLUEWEED
fl. ⅝″

HOUND'S
TONGUE   fl. ¼″

CORN GROMWELL
fl. ¼″

HOARY PUCCOON
fl. ⅜″

PUCCOON   fl. ½″

GROMWELL   fl. ½″

PLANTAIN fl. arr. 5"

BUCKHORN fl. arr. 5"

ENGLISH PLANTAIN
fl. arr. 2½"

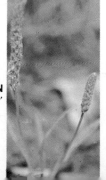

SAND PLANTAIN
fl. arr. 2"

PLANTAIN FAMILY

PLANTAIN, *Plantago rugelii* / June–Aug. / 3"–12" / Disturbed ground (*P. major*: leafstalks not red at base, disturbed ground; *P. cordata*: leaves feather-veined, spike stalk hollow, rivers and swamps)

BUCKHORN, *P. aristata* / June–Nov. / ½"–10" / Dry open disturbed ground / Leaves slender, not woolly

ENGLISH PLANTAIN, *P. lanceolata* / June–Aug. / 1½'–2' / Disturbed ground / Leaves long, slender—like a coarse grass (*P. indica*: stems opposite-leaved and branched; spikes short, rounded)

SAND PLANTAIN, *P. patagonica* / June–Aug. / ½"– 10" / Dry sand prairies and roadsides / Plant white-woolly; leaves slender

DOGBANE FAMILY

DOGBANE, *Apocynum androsaemifolium* /June–Aug. / 8"–24" / Edges and openings of dry woods and forests / Flowers pink-striped, drooping

INDIAN HEMP, *A. sibiricum* / June–Aug. / 1'–5' / Meadows, shores, woods edges / Forms dense patches

WATERLEAF FAMILY

WATERLEAF, *Hydrophyllum virginianum* / May–June / 6"–30" / Wet to medium woods and forests / Lavender or white; basal leaves feather-divided (*Ellisia nyctelea*: flowers solitary along stem; leaves opposite, feather-lobed or divided; plant weakly reclining)

HAIRY WATERLEAF, *H. appendiculatum* / May–July / 1'–4' / Wet to medium woods and forests / Leaves velvety, maple-like

TOMATO FAMILY

NIGHTSHADE, *Solanum dulcamara* / June–Sept. / Trailing or twining to 8' / Moist disturbed ground, open woods, cliffs, marshes / Berry poisonous to eat, red

BLACK NIGHTSHADE, *S. nigrum* / June–Oct. / 4"–24" / Disturbed ground / Ripe berry black, poisonous to eat

DOGBANE fl. ⅜″

INDIAN HEMP fl. ¼″

WATERLEAF fl. ¾″

HAIRY WATERLEAF fl. ¾″

BLACK NIGHTSHADE fl. 3/16″

NIGHTSHADE fl. ¼″

HORSE NETTLE fl. ¾"

GROUND CHERRY fl. ⅝"

**LARGEFLOWER
GROUND CHERRY** fl. 1⅜"

CLAMMY GROUND CHERRY fl. 1"

JIMSON WEED fl. 1¾"

MORNING GLORY
fl. 4"

FIELD BINDWEED fl. ¾″

ERECT BINDWEED fl. 2½″

HEDGE BINDWEED fl. 2½″

DODDER

right
⅛″ **flower**

TOMATO FAMILY

HORSE NETTLE, *Solanum carolinense* / June–Sept. / 20″–40″ / Disturbed ground, often sandy / Lavender to white; berry yellow; plant spiny

GROUND CHERRY, *Physalis virginiana* / July–Aug. / ½′–2′ / Dry prairies and open woods
(APPLE OF PERU, *Nicandra physalodes:* flower pale blue; plant 4″–50″; disturbed ground) The fruit of the Ground Cherry is a round berry in a papery husk

LARGEFLOWER GROUND CHERRY, *P. grandiflora* / June–Aug. / 1′–3′ / Dry sandy or rocky soil—north

CLAMMY GROUND CHERRY, *P. heterophylla* / June–Sept. / 8″–36″ / Dry woods and prairies, sandy soil / Plant sticky-hairy

JIMSON WEED, *Datura stramonium* / July–Sept. / 2′–3′ / Sunny disturbed ground / Flowers open about noon; plant rank-smelling; fruit a dry, spiny pod

MORNING GLORY FAMILY

MORNING GLORY, *Ipomaea pandurata* / June–Sept. / Twining or trailing to 18′ / Dry woods—south (Many annual Morning Glories escape to disturbed ground)

FIELD BINDWEED, *Convolvulus arvensis* / June–Sept. / Twining or trailing to 3′ / Disturbed ground

ERECT BINDWEED, *C. spithameus* / June–July / 3″– 20″, not twining / Sandy or rocky soil, dry open woods and forests

HEDGE BINDWEED, *C. sepium* / June–Sept. / Twining or trailing to 10′ / Disturbed ground, prairies, wet meadows / Pink or white

DODDER, *Cuscuta gronovii* / July–Oct. / 6″–72″ / Leafless parasite twining on moist-ground plants

WILD PETUNIA   fl. 1¼"

JUSTICIA   fl. ½"

HEDGE-NETTLE   fl. ⅜"

GERMANDER   fl. ⅜"

ACANTHUS FAMILY

WILD PETUNIA, *Ruellia humilis* / June–Aug. / 3"–24" / Dry prairies and woods—south

JUSTICIA, *Justicia americana* / June–Oct. / 8"–40" / Shallow water, mud—south / Violet to white

MINT FAMILY

HEDGE-NETTLE, *Stachys palustris* / July–Aug. / 8"–40" / Wet meadows, prairies, woods, shores

GERMANDER, *Teucrium canadense* / July–Aug. / 12"–40" / Medium to wet woods and prairies / Large lower lip fades to tan

SIDEFLOWER SKULLCAP, *Scutellaria lateriflora* / June–Sept. / 4"–36" / Wet woods, shores, shallow water / All Skullcaps: helmet-shaped sepal-tubes

SMALL SKULLCAP, *S. leonardi* / June–July / 3"–8" / Dry prairies and open woods/ Underground runners

SKULLCAP, *S. ovata* / June–July / 12"–28" / Medium to dry woods and forests

MARSH SKULLCAP, *S. galericulata* / June–Aug. / 6"– 32" / Shores, marshes, wet meadows

MOTHERWORT, *Leonurus cardiaca* / June–Aug. / 18"–60" / Disturbed ground, sun or shade / Flower fuzzy; sepals spiny

BLEPHILIA, *Blephilia ciliata* / June–Aug. / 5"–36" / Medium to dry woods, dry to wet prairies

DOWNY BLEPHILLIA, *B. hirsuta* / May–Aug. / 16"–32" / Wet to medium woods

DRAGONHEAD, *Physostegia virginiana* / Aug.–Oct. / 2'–3' / Medium to wet woods and prairies / Plant not hairy

SIDEFLOWER SKULLCAP   fl. ⅜"

SKULLCAP   fl. ¾"

SMALL SKULLCAP
fl. ⅜"

MARSH SKULLCAP
fl. ⅝"

DOWNY BLEPHILIA   fl. arr. 2"

BLEPHILIA   fl. arr. 1¾"

MOTHERWORT   fl. ¼"

DRAGONHEAD   fl. 1"

HEMP-NETTLE fl. ¾″

WILD BASIL fl. 3/16″

HEAL ALL fl. arr. 1¼″

GIANT-HYSSOP fl. arr. 6″

GILL OVER THE GROUND fl. ¼″

BUGLEWEED fl. ⅝″

HENBIT fl. ⅝″

MINT FAMILY

HEMP-NETTLE, *Galeopsis tretrahit* / June/Sept. / 6″–36″ / Disturbed ground, more abundant north / Plant bristly-hairy

WILD BASIL, *Satureja vulgaris* / July–Sept. / 5″–24″ / Woods and forest openings

GIANT-HYSSOP, *Agastache scrophulariaefolia* / Aug.–Sept. / 3′–5′ / Dry to wet woods
(*A. foeniculum:* flowers blue; leaves anise-scented, white beneath; prairies, inland sands. *A. nepetoides:* flowers greenish-yellow; woods. *Dracocephalum parviflorum:* purple flowers nearly hidden by spiny-toothed leaves; rocky limy soil—north)

SAVORY   fl. 3/16″

HEAL ALL, *Prunella vulgaris* / June–Oct. / 1″–20″ / Disturbed ground, sun or shade

BUGLEWEED, *Ajuga genevensis* / April–June / 4″–12″ / Disturbed ground, sun or shade / Forms patches; lower lip 3-parted

HENBIT, *Lamium amplexicaule* / April–Oct. / 2″–14″ / Disturbed ground / Lower leaves stalked, upper ones clasp stem

The remaining Mint family members have **AROMATIC ODOR** when crushed

GRASSLEAF PENNYROYAL
fl. ⅛″

GILL OVER THE GROUND, *Glecoma hederacea* / April–June / 1″–8″, creeping to 4′ / Medium to wet woods, disturbed ground

SAVORY, *Satureja glabella* / June–Aug. / 2″–8″ / Damp limy places, particularly Great Lakes beaches / Plant without hairs

GRASSLEAF PENNYROYAL, *Hedeoma hispida* / May–Aug. / 1″–16″ / Inland sands, dry open soil / Leaves very narrow
(AMERICAN PENNYROYAL, *H. pulegioides:* leaves oval, stalked)

CATNIP, *Nepeta cataria* / July–Aug. / 18″–40″ / Disturbed ground, dry open woods / Flower purple-dotted

CATNIP   fl. arr. 2″

MOUNTAIN MINT fl. arr. 3″

MINT fl. 1/8″

SPEARMINT fl. arr. 2″

WATER HOREHOUND fl. 1/8″

HORSE BALM fl. 5/8″

BERGAMOT fl. arr. 1½″

DOTTED MINT fl. 1/4″

LOPSEED fl. 1/4″

MINT FAMILY

MOUNTAIN MINT, *Pycnanthemum virginianum* / July–Sept. / 1′–3′ / Wet meadows, prairies, woods / Flower arrangement flat-topped

MINT, *Mentha arvensis* / July–Sept. / 4″–32″ / Wet meadows, shores, streambanks / Leaves fuzzy-hairy; flowers up-and-down stem

SPEARMINT, *M. spicata* / July–Oct. / 1′–3′ / Wet meadows, streambanks / Flowers at the end of stem
(PEPPERMINT, *M. piperita:* leaf stalked)

WATER HOREHOUND, *Lycopus uniflorus* / July–Sept. / 2″–32″ / Wet meadows, marshes, streambanks, damp cliffs—usually near water
(*L. americanus:* leaf somewhat lobed)

HORSE BALM, *Collinsonia canadensis* / July–Sept. / 20″–40″ / Woods

BERGAMOT, *Monarda fistulosa* / July–Sept. / 2′–4′ / Dry woods, prairies, fields
(OSWEGO TEA, *M. didyma:* flowers brilliant red; moist woods—east, gardens)

DOTTED MINT, *M. punctata* / July–Sept. / 6″–36″ / Inland sands / Upper leaves white-green to lavender; flowers yellowish with purple spots

LOPSEED FAMILY

LOPSEED, *Phryma leptostachya* / July–Aug. / 1′–3′ / Dry to medium woods and forests / Fruit 3-barbed stick-tight, pointing down

SNAPDRAGON FAMILY

BEARDTONGUE, *Penstemon grandiflorus* / May–June / 12″–40″ / Dry prairies, inland sands / Leaf rounded, blue–green; hairless (Penstemons have 1 brush-like stamen)

PENSTEMON, *P. digitalis* / May–July / 1′–4′ / Medium prairies, fields, open woods / Leaf finely-toothed
(*P. tubaeflorus:* flowers sticky-hairy inside *and* out; leaves untoothed)

SLENDER PENSTEMON, *P. gracilis* / June–July / 8″–24″ / Dry prairies, inland sands, open woods, disturbed ground / Leaves very narrow
(*P. hirsutus:* stem conspicuously hairy; flower tube closed at mouth)

BEARDTONGUE   fl. 1¾″

PENSTEMON   fl. 1″

SLENDER PENSTEMON fl. ¾″

TURTLE HEAD fl. 1⅛"

HEDGE-HYSSOP fl. ½"

GOLD HEDGE-HYSSOP fl. ¾"

COW WHEAT fl. ⅜"

MONKEY FLOWER fl. 1"

YELLOW MONKEY FLOWER fl. ½"

FIGWORT fl. ⅜"

PINK GERARDIA fl. 1¾"

GERARDIA fl. ½"

SNAPDRAGON FAMILY

TURTLE HEAD, *Chelone glabra* / July–Sept. / 2'–3' / Streambanks, wet meadows, forests and woods /

HEDGE-HYSSOP, *Gratiola neglecta* / May–Oct. / 2"–12" / Wet soil, mud, streambanks

GOLD HEDGE-HYSSOP, *G. aurea* / July–Sept. / 1"–12" / Peaty or sandy shores, usually acid soil

MONKEY FLOWER, *Mimulus ringens* / July–Sept. / 6"–40" / Wet meadows, shores, marshes, streambanks

CORN SPEEDWELL fl. 1/8"

YELLOW MONKEY FLOWER, *M. glabratus* / June–Aug. / 1"–20" / Shallow water of springs and brooks

SPEEDWELL fl. 1/8"

COW WHEAT, *Melampyrum lineare* / July–Aug. / 3"–16" / Bogs, damp rocky, peaty, or sandy forests / Flowers in pairs; seeds few

FIGWORT, *Scrophularia lanceolata* / May–Aug. / 2'–6' / Meadows, fields, woods edges, rocky soil / Stem square, flat-sided

GERARDIA, *Gerardia tenuifolia* / Aug.–Sept. / 3"–26" / Shores, damp meadows, sand, peat, or marl / Flowers long-stalked

PINK GERARDIA, *G. purpurea* / Aug.–Sept. / 3"–36" / Same habitats / Flowers very short-stalked (Other Gerardias: dry prairies, open woods.)

AMERICAN BROOKLIME fl. 3/16"

CORN SPEEDWELL, *Veronica arvensis* / April–Aug. / 1"–12" / Disturbed ground / All Veronicas 4-petaled (*V. serpyllifolia:* flower pale blue with violet lines; creeps to form leafy mats)

WATER SPEEDWELL fl. 3/16"

SPEEDWELL, *V. peregrina* / May–Aug. / 1"–12" / Disturbed ground / Upper leaves alternate

WATER SPEEDWELL, *V. catenata* / June–Oct. / 4"–10", trailing / Shores, streams / Leaves stalkless

AMERICAN BROOKLIME, *V. americana* / June–Oct. / 4"–10", trailing / Shores, streambanks, shallow water / Leaves succulent, stalked

SNAPDRAGON FAMILY

FALSE FOXGLOVE, *Aureolaria grandiflora* / July–Sept. / 20"–40" / Dry open woods / Lower leaves feather-lobed, upper often only toothed
(*A. pedicularia:* all leaves lobed, plant sticky-hairy; *A. flava:* plant not hairy; *A. virginica:* seed capsule fuzzy)

CULVER'S ROOT, *Veronicastrum virginicum* / July–Aug. / 3'–6' / Wet to dry open woods, wet prairies / Leaves whorled

MOTH MULLEIN, *Verbascum blattaria* / June–Oct. / 1'–3' / Disturbed ground / Yellow, white; stamens with violet woolly hairs

MULLEIN, *V. thapsus* / June–Sept. / 1'–6' / Disturbed ground / Leaves very woolly with branching hairs; stem winged
(*V. phlomoides:* stem not winged; flower 1½")

BUTTER AND EGGS, *Linaria vulgaris* / May–Sept. / 1'–3' / Disturbed ground, sandy soil / Plant without hairs
(Notice the nectar spur, a character of all Linarias including OLD-FIELD TOAD-FLAX, *L. canadensis:* ⅜" flowers purple, basal rosettes of tiny succulent leaves that turn yellow, dry sandy soil; *L. minus:* ⅜" flowers lilac, fruit and plant sticky-hairy, disturbed ground)

WOOD BETONY, *Pedicularis canadensis* / April–June / 4"–16" / Medium to dry prairies, forests, woods / Flowers curve to the side, yellow to maroon

LOUSEWORT, *P. lanceolata* / Aug.–Sept. / 1'–4' / Limy wet meadows and prairies, marshes, shores / Flowers curve to the side

KITTEN TAIL, *Besseya bullii* / May / 8"–16" / Dry prairies, open woods, sand or gravel soil / Leaves rounded, flat on ground

INDIAN PAINT BRUSH, *Castilleja coccinea* / May–Aug. / 5"–24" / Wet to dry meadows and prairies, wet sand / Flowers red to yellow

DOWNY PAINTED CUP, *C. sessiliflora* / May–July / 5"–12" / Dry limy prairies and hillsides—west

FALSE FOXGLOVE   fl. 1¼"

CULVER'S ROOT   fl. arr. 6"

BUTTER AND EGGS   fl. ¾"

KITTEN TAIL
fl. arr. 6"

MOTH MULLEIN   fl. 1¼"

MULLEIN   fl. spike 20"

WOOD BETONY   fl. ⅞"

LOUSEWORT   fl. ⅞"

INDIAN PAINT BRUSH   fl. 1"

DOWNY PAINTED CUP   fl. 1½"

GREAT BLUE LOBELIA fl. ⅞"

WATER LOBELIA fl. ¾"

CARDINAL FLOWER fl. 1"

INDIAN TOBACCO
fl. ⅜"

PALE
SPIKE LOBELIA fl. ½"

BROOK LOBELIA
fl. ½"

BUTTERWORT fl. ⅞"

BLADDERWORT fl. ⅝"

HORNED
BLADDERWORT fl. ⅝"

SQUAW ROOT   fl. ⅝"

102—103

LOBELIA FAMILY

GREAT BLUE LOBELIA, *Lobelia siphilitica* / Aug.–Sept. / 1'–4' / Wet meadows, streambanks, shores

WATER LOBELIA, L. dortmanna / July–Sept. / 4"–40" / Shallow water —north / Leaves slender, succulent

CARDINAL FLOWER, L. cardinalis / July–Sept. / 2'–5' / Damp meadows, wet open woods, streambanks

BROOK LOBELIA, L. kalmii / July–Sept. / 4"–16" / Limy bogs, shores, wet meadows / Flowers white-centered; leaves very narrow

INDIAN TOBACCO, L. inflata / July–Oct. / 1'–3' / Disturbed woods, forests / Rounded seed capsules

PALE SPIKE LOBELIA, L. spicata / June–July / 8"–40" / Wet to dry meadows, prairies, open woods

BLADDERWORT FAMILY

BUTTERWORT, Pinguicula vulgaris / June–July / 2"–6" / Bogs, rocky shores, wet meadows—north / Leaves sticky above, edges uprolled

BLADDERWORT, Utricularia intermedia / July–Aug. / Flowering stalk 3"–10" / Nectar spur short; underwater leaves finely-divided (Purple species: *U. purpurea, U. resupinata*)

HORNED BLADDERWORT, U. cornuta / July–Aug. / Flowering stalk 3"–14" / Bogs, wet shores / Nectar spur long; leaves hidden in mud

BROOMRAPE FAMILY

SQUAW ROOT, Conopholis americana / May–July / 2"–8" / Dry to medium woods, especially under oaks / Plant resembles a pine cone

BEECH DROPS, Epifagus virginiana / Aug.–Oct. / 4"–20" / Medium woods, on beech roots

SAND CANCER-ROOT, Orobanche fasciculata / May–Aug. / 2"–6" / Dry prairies, inland sands (*O. ludoviciana:* flowers in spikes)

CANCER-ROOT, O. uniflora / May–June / 2"–8" / Dry to wet woods or open ground / Violet to white

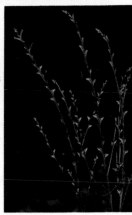

BEECH DROPS fl. ⅜"

SAND CANCER-ROOT   fl. ⅞"

CANCER-ROOT   fl. ¾"

DANDELION hd. 1″

PRAIRIE DANDELION hd. 1″

DWARF DANDELION hd. ⅜″

CYNTHIA hd. ¾″

MEADOW GOAT'S BEARD fl. hd. 1¼″
seed head below, 3″

DAISY FAMILY, Dandelion tribe

DANDELION, *Taraxacum officinale* /
April–Nov. / 2″- 12″ / Disturbed
ground—everywhere

HAWK'S BEARD   hd. 1″

**PRAIRIE DANDELION, *Microseris
cuspidata*** / April–May / 3″–12″ /
Dry prairies, rocky hills / Leaf wavy-
edged, midrib silvery

CYNTHIA, *Krigia biflora* / May–Aug.
/ 8″–32″ / Wet to medium meadows,
prairies, open woods / 1 (occ. to
3) smooth rounded leaf on stem

**ORANGE
HAWKWEED
hd. ½″**

DWARF DANDELION, *K. virginica* /
April–June / 2″–16″ / Dry open
sandy soil / Leaves small, all basal

**MEADOW GOAT'S BEARD, *Trago-
pogon dubius*** / May–July / 1′–3′ /
Disturbed ground, roadsides /
Grass-like leaves sheathe stem;
flowers open only in morning sun
(*T. porrifolius:* flowers purple)

**HAWKWEED
hd. ⅝″**

HAWK'S BEARD, *Crepis tectorum*
/ June–Oct. / 8″–20″ / Sandy disturbed ground / Leaves often feather-lobed; plant not hairy, unlike the
Hawkweeds

**ORANGE HAWKWEED, *Hieracium
aurantiacum*** / June–Oct. / Disturbed
ground / Dense patches or mats;
plant very hairy; leaves basal
(*H. pratense:* yellow)

HAWKWEED, *H. canadense* / July–
Oct. / ½′–5′ / Dry woods, forests
and prairies, cliffs, roadsides /
Leaves toothed or sharp-lobed; stem
leafy
(*H. scabrum:* leaves rounded, plant
bristly-hairy, dry open woods and forests)

**LONGHAIR HAWKWEED, *H. longipi-
lum*** / July–Aug. / 2′–5′ / Dry prairies and fields, sandy soil / Leaves
mostly basal; hairs on stem over ½″

LONGHAIR HAWKWEED   hd. ½″

PRICKLY WILD LETTUCE hd. ⅜"

WILD LETTUCE hd. ⅜"

RATTLESNAKE ROOT
hd. ½"

LION'S FOOT hd. ⅝"

SOW THISTLE hd. 1¼"

CHICORY hd. 1¼"

DAISY FAMILY, Dandelion tribe

PRICKLY WILD LETTUCE, *Lactuca scariola* / July–Sept. / 1'–5' / Disturbed ground / Leaves blue-green, midrib bristly beneath

WILD LETTUCE, *L. canadensis* / July–Sept. / 1'–8' / Meadows, fields, open woods / Yellow to purple (*L. ludoviciana:* leaves blue-green, spiny-margined; *L. pulchella* and *L. floridana:* heads blue. All Lettuces: leaves variable, often lobed, early clustered heads become loosely spread)

RATTLESNAKE ROOT, *Prenanthes racemosa* / Aug.–Oct. / 1'–5' / Wet to dry prairies, meadows, streambanks / Leaves blue-green, rounded; back of heads hairy

LION'S FOOT, *P. alba* / Aug.–Oct. / 1½'–5' / Wet to dry woods and forests, rocky places / Basal leaves triangular, others variable; back of heads smooth (All of *Prenanthes* have drooping, cylindric heads.)

SOW THISTLE, *Sonchus arvensis* / July–Oct. / 16"–60" / Disturbed ground / Forms patches (*S. oleraceus:* grows singly; leaves coarse, clasp stem)

CHICORY, *Cichorium intybus* / July –Oct. / 1'–6' / Roadsides, disturbed ground / Leaves mostly basal, lobed; flowers may also be white

DAISY FAMILY, Thistle-like tribes

CENTAUREA, *Centaurea maculosa* / June–Oct. / 1'–4' / Disturbed ground, often limy / Stem wiry, much-branching

STAR THISTLE, *C. scabiosa* / June–Oct. / 1'–3' / Disturbed ground, often sandy / Plant leafy at base only / *Centaureas* may also be white

BURDOCK, *Arctium minus* / July–Oct. / 1½'–5' / Disturbed ground, sun or shade / Heads sharp-hooked, clustered; basal leaves large, fuzzy (*A. lappa:* heads long-stalked. True Docks, *Rumex*, hairless. See also Cocklebur and Prairie Dock)

CENTAUREA   hd. 1"

STAR THISTLE   hd. 1"

BURDOCK   hd. ¾"

SWAMP THISTLE   hd. 1½″

BULL THISTLE   hd. 1½″

BEACH THISTLE   hd. 1½″

FIELD THISTLE   hd. 1½″

CANADA THISTLE   hd. ½″

BRISTLY THISTLE   hd. ¾″

NODDING THISTLE
hd. 2″

DAISY FAMILY, Thistle-like tribes

SWAMP THISTLE, *Cirsium muticum* / July–Oct. / 2'–7' / Wet woods and meadows, streambanks / Back of flower heads sticky—not prickly

BULL THISTLE, *C. vulgare* / June–Oct. / 2'–6' / Disturbed ground, pastures, marshes / Leaves bristly above; stem prickly-winged
(*C. palustre,* heads 1/2", wet forests; SCOTCH THISTLE, *Onopordum acanthium,* plant cottony-woolly, disturbed ground)

BEACH THISTLE, *C. pitcheri* / May –Sept. / 2'–4' / Sand beaches of the Great Lakes / Plant white-woolly

FIELD THISTLE, *C. discolor* / July–Oct. / 3'–7' / Prairies, fields / Flowers pale; leaves whitened beneath
(*C. hillii:* leaves not whitened; heads 2"; plant 1'–3'; June–Aug.)

CANADA THISTLE, *C. arvense* / June–Oct. / 1'–4' / Disturbed ground, pastures / Heads numerous; forms extensive patches

BRISTLY THISTLE, *Carduus acanthoides* / July–Oct. / 1'–4' / Disturbed ground, pastures / Stem spiny-winged

NODDING THISTLE, *C. nutans* / June–Oct. / 1'–7' / Heads tipped to the side; stem spiny-winged

(Blazing Stars start blooming at top. Bracts covering back of flower are species clue)

BLAZING STAR, *Liatris aspera* / Aug.–Oct. / 16"–48" / Dry to medium prairies, dry woods, inland sands / Bracts cup-shaped, white to pink on edges
(*L. ligulistylis:* bracts flat, red or purple; top head often larger; prairies)

DWARF BLAZING STAR, *L. cylindracea* / July–Sept. / 6"–24" / Dry rocky or sandy prairies / Bracts rounded to pointed, lying flat
(*L. punctata:* heads sparsely flowered; leaves minutely dotted; prairies—west)

GAYFEATHER, *L. pycnostachya* / July–Sept. / 2'–4' / Medium to wet meadows and prairies / Bracts reddish, sharp-pointed, fuzzy, bent out
(*L. spicata:* bracts lying flat, smooth)

BLAZING STAR
hd. 1"

DWARF BLAZING STAR   hd. ⅝"

GAYFEATHER   hd. ½"

DAISY FAMILY, Thistle-like tribes

WHITE SNAKEROOT, *Eupatorium rugosum* / July–Oct. / 1'–5' / Medium woods, rocky soil, disturbed soil / Leaves opposite, stalked (*E. altissimum,* leaves slender; *E. sessilifolium:* leaves unstalked)

BONESET, *E. perfoliatum* / July–Oct. / Wet meadows and prairies, rocky shores / Veiny paired leaves joined around stem

JOE-PYE WEED, *E. maculatum* / July–Sept. / 2'–7' / Damp meadows, marshes, shores / Leaves whorled (*E. purpureum:* flowers in open domes; dry to medium woods, clearings)

IRONWEED, *Vernonia fasciculata* / July–Sept. / 2'–6' / Wet meadows, pastures, prairies / Leaves alternate, numerous, without hairs

FALSE BONESET, *Kuhnia eupatorioides* / Aug.–Oct. / 1'–4' / Dry limy prairies and open woods / Leaves alternate, variable in width

DAISY FAMILY, Daisy tribe

DAISY, *Chrysanthemum leucanthemum* / June–Aug. / 8"–24" / Disturbed ground, pastures, fields / Basal leaves round-toothed or lobed

CHAMOMILE, *Anthemis cotula* / May–Oct. / 4"–24" / Disturbed ground, common in farmyards / Leaves very finely-cut, ill-scented

TANSY, *Tanacetum vulgare* / July–Oct. / Fields, roadsides, disturbed ground / Flower head has no rays

PINEAPPLE WEED, *Matricaria matricarioides* / May–Sept. / 1"–18" / Disturbed ground / No rays; crushed leaves have pineapple odor

YARROW, *Achillea millefolium* / June–Oct. / 8"–40" / Fields, roadsides, disturbed ground / Occ. pink; leaves finely-divided, with acrid odor

DAISY FAMILY, Sunflower tribe

WILD QUININE, *Parthenium integrifolium* / July–Sept. / 20"–40" / Medium to dry prairies and woods / Heads have 5 tiny rounded rays; basal leaves coarse, undivided

WHITE SNAKEROOT
fl. arr. 3" wide

BONESET fl. arr. 6" wide

JOE-PYE WEED fl. arr. 7" wide

IRONWEED   fl. arr. 8" wide

DAISY   hd. 2"

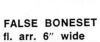

FALSE BONESET
fl. arr. 6" wide

CHAMOMILE
hd. ⅝"

TANSY
hd. ¼"

WILD QUININE   hd. ¼"

YARROW   hd. 3/16"

PINEAPPLE
WEED   hd. ⅜."

WHITE SAGE
fl. arr. 6"

WORMWOOD
fl. arr. 14"

MARSH-ELDER   fl. arr. 10"

COCKLEBUR   bur 1"

RAGWEED   fl. arr. 6"

GIANT RAGWEED   fl. arr. 10"

QUICKWEED   hd. 3/16"

DAISY FAMILY, Daisy tribe

WHITE SAGE, *Artemisia ludoviciana* / July–Oct. / 1'–3' / Dry prairies, disturbed ground / Forms patches; leaves white both sides
(SAWTOOTH SAGE, *A. serrata*: 3'–6'; leaves white only beneath; streambanks, prairies)
'Sage' may be used for western sagebrushes and several garden *Artemisias*; blue—flowered *Salvias* of Mint family.

WORMWOOD, *A. campestris* / July–Oct. / 1'–4' / Dry sandy prairies and open woods, beaches / Grows singly; leaves finely divided

TALL BEGGARTICKS   hd. ½"

DAISY FAMILY, Sunflower tribe

MARSH-ELDER, *Iva xanthifolia* / Aug.–Oct. / 1½'–6' / Damp disturbed ground / Leaves undivided

RAGWEED, *Ambrosia artemisiifolia* / Aug.–Sept. / 3"–40" / Disturbed ground / Leaves feather-lobed, opposite below, alternate above
(SAND RAGWEED, *A. psilostachya*: leaves obtusely-lobed; forms patches)

GIANT RAGWEED, *A. trifida* / Aug.–Sept. / 2'–9' / Disturbed ground, usually damp / leaves 3–5 lobed

COCKLEBUR, *Xanthium strumarium* / July–Sept. / 8"–80" / Shores, disturbed ground / Leaves rough, occ. lobed; pointed prickly burs clustered on stem

QUICKWEED, *Galinsoga ciliata* / June–Oct. / 8"–28" / Disturbed ground, sun or shade

BEGGARTICKS, *Bidens cernua* / Aug.–Oct. / 1"–60" / Wet meadows, marshes, shores / Flower heads tip sideways
(*B. beckii*: submerged leaves finely-divided; shallow water)

TALL BEGGARTICKS, *B. vulgata* / Aug.–Oct. / 4"–60" / Disturbed ground, wet or dry / Flower heads have no rays

BUR MARIGOLD, *B. coronata* / Aug.–Oct. / 6"–60" / Wet meadows, marshes, shores

(Seeds of all Bidens have 2 or 4 barbed spines)

BEGGARTICKS   hd. 1"

BUR MARIGOLD   hd. 1½"

PURPLE CONEFLOWER rays 2″

PRAIRIE CONEFLOWER rays 2″

TALL CONEFLOWER hd. 1¾″

BRANCHED CONEFLOWER hd. 1¼″

BLACK-EYED SUSAN hd. 1½″

SNEEZEWEED   hd. 1"

COREOPSIS   hd. 1½"

below, back view

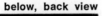

DAISY FAMILY, Sunflower tribe

PURPLE CONEFLOWER, *Echinacea purpurea* / June–Oct. / 2'–5' / Prairies, dry open woods—south / Center of head spiny
(PALE CONEFLOWER, *E. pallida:* rays and leaves very slender, long; dry to medium prairies)

PRAIRIE CONEFLOWER, *Ratibida pinnata* / June–Aug. / 1½'–4' / Moist to dry prairies, open woods / Rays always hang down; lower leaves feather-divided
(*R. columnifera:* disc a cylindrical column; dry disturbed ground, prairies)

TALL CONEFLOWER, *Rudbeckia laciniata* / July–Sept. / 2'–10' / Moist woods and forest edges, shores, streambanks / Forms patches; stems leafy

BRANCHED CONEFLOWER, *R. triloba* / July–Oct. / 1'–5' / Disturbed ground, open woods / Plant bushy; lower leaves 3–7 lobed
(*R. subtomentosa:* 2'–6'; leafy basal offshoots; wet meadows and prairies)

BLACK-EYED SUSAN, *R. hirta* / June–Oct. / 1'–3' / Meadows, prairies, open woods / Whole plant conspicuously hairy

SNEEZEWEED, *Helenium autumnale* / Aug.–Oct. / 2'–5' / Moist meadows, shores, streambanks / Stem winged; rays appear squarely cut

COREOPSIS, *Coreopsis palmata* / June–Aug. / 2'–3' / Dry to medium prairies, dry open woods / Forms patches; leaves deeply 3-lobed; notice, in back view picture, the plastic-like yellow-green bracts beneath the head; plant hairless
(*C. lanceolata:* slender unlobed leaves, mostly basal; dry sand or rocks; May–Aug.; plant smooth or hairy)

DAISY FAMILY, Sunflower tribe

LEAFCUP, *Polymnia canadensis* / June–Sept. / 1'–4' / Damp woods and forests, cliffs / Head often without rays; leaf-pairs form cup around stem; plant sticky-hairy

CUP PLANT, *Silphium perfoliatum* / July–Sept. / 3'–8' / Wet prairies, woods, streambanks / Leaf-pairs form cups around stem which is square

LEAFCUP   hd. ⅜"

CUP PLANT   hd. 2½"

COMPASS PLANT, *S. laciniatum* / June–Sept. / 3'–7' / Medium to dry prairies / Stem bristly-hairy

PRAIRIE DOCK, *S. terebinthinaceum* July–Sept. / 2'–10' / Wet to medium prairies, in deep soil

ROSINWEED, *S. integrifolium* / July –Sept. / 2'–6' / Dry sandy to medium prairies and meadows, roadsides

(Note: Silphium seeds are in a ring AROUND disc; Sunflower's COVER disc)

OX EYE, *Heliopsis helianthoides* / July–Sept. / 2'–5' / Dry woods, dry to wet prairies, disturbed ground / Forms clumps; rays do not fall off as they do in *Helianthus*

(Note: All our species of *Helianthus* form patches. The ANNUAL SUNFLOWER, a common garden escape, does not)

JERUSALEM ARTICHOKE, *Helianthus tuberosus* / Aug.–Oct. / 4'–10' / Disturbed ground, usually moist / Upper leaves alternate; leaves broad, with winged stalks; stem fuzzy

NAKED SUNFLOWER, *H. occidentalis* / July–Sept./ 2'–6' / Dry prairies and open woods, sand, rocks / Leaves mostly basal

COMPASS PLANT   hd. 3"

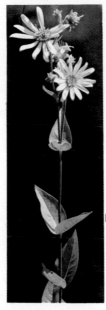

PRAIRIE DOCK   hd. 2½"

ROSINWEED   hd. 2½"

OX EYE   hd. 2½"

JERUSALEM ARTICHOKE   hd. 2½"

NAKED SUNFLOWER   hd. 2"

WOODLAND SUNFLOWER

hd. 2½"

SAWTOOTH SUNFLOWER

SHOWY SUNFLOWER   hd. 3"

DAISY FAMILY, Sunflower tribe

WOODLAND SUNFLOWER, *Helianthus strumosus* / July–Sept. / 2'–6' / Dry to wet woods / Leaf very rough, on winged stalk

SAWTOOTH SUNFLOWER, *H. grosseserratus* / July–Oct. / 2'–10' / Wet prairies, marsh edges / Stem very leafy to top, smooth; upper leaves alternate, sparsely toothed
(*H. giganteus:* stem white-hairy; damp meadows, roadsides, forest clearings)

SHOWY SUNFLOWER, *H. laetiflorus* / Aug.–Sept. / 1½'–8' / Prairies / Disc often dark; leaf rough, stiff

ELECAMPANE, *Inula helenium* / June–Aug. / 3'–6' / Disturbed ground, woods and forest clearings / Plant coarse; rays numerous

DAISY FAMILY, Aster-like tribes

GUMWEED, *Grindelia squarrosa* / Aug.–Sept. / 4"–40" / Dry disturbed ground / Flower heads very sticky

FLEABANE, *Erigeron philadelphicus* / May–Aug. / 4"–36" / Wet woods, shores, meadows / Rays over 100; young heads droop; white or pink

ROBIN'S FLEABANE, *E. pulchellus* / May–June / 6"–18" / Dry open woods / In patches; rays white or lavender; leaves fuzzy, mostly basal

DAISY FLEABANE, *E. strigosus* / June–Sept. 1'–3' / Fields, roadsides, disturbed ground, open woods
(*E. annuus:* leaf bristly-hairy, toothed)

HORSEWEED, *Conyza canadensis* / July–Oct. / 3"–60" / Disturbed ground / Leaves slender, long; stem very leafy

RAGWORT, *Senecio pauperculus* / May–Aug. / 4"–24" / Wet rocks, meadows, prairies / Patches; basal leaves slender-oval
(*S. plattensis:* white-woolly; dry prairies)

GOLDEN GROUNDSEL, *S. aureus* / May–Aug. / 6"–36" / Wet woods and forests, meadows, streambanks / Basal leaves heart-shaped, rounded
(*S. vulgaris:* rayless, leaves fleshy, wavy)

RAGWORT   hd. ½″

ELECAMPANE   hd. 2½″

**GOLDEN GROUNDSEL
hd. ½″**

GUMWEED   hd. 1″

FLEABANE   hd. ¾″

HORSEWEED   hd. 3/16″

**DAISY FLEABANE
hd. ½″**

**ROBIN'S FLEABANE
hd. ¾″**

LARGELEAF ASTER hd. ¾″

FLAT-TOP ASTER hd. ½″

ARROW ASTER hd. ½″

SMOOTH ASTER hd. ¾″

WHITE ASTER hd. ½″

MARSH ASTER hd. ½″

HEATH ASTER hd. ⅜"

FROST ASTER hd. ⅝"

BOG ASTER hd. ⅞"

DAISY FAMILY, Aster-like tribes

LARGELEAF ASTER, *Aster macrophyllus* / July–Oct. / 1'–3½' / Woods and forests / Patches of large heart-shaped leaves

FLAT-TOP ASTER, *A. umbellatus* / July–Sept. / 1'–7' / Dry to wet meadows, wet woods, forest edges / Stems leafy; rays look unkempt

WHITE ASTER, *A. ptarmicoides* / July–Sept. / 3"–24" / Dry prairies, inland sands, cliffs / Flat-topped; in clumps; leaves slender, rigid

ARROW ASTER, *A. sagittifolius* / Aug.–Oct. / 1'–4' / Dry to medium woods, disturbed ground, streambanks / Pyramid-topped, white to blue; in clumps; lower leaves arrow-shaped with winged stalks
(*A. ciliolatus, A. cordifolius*, arrowleaf Asters of forest clearings: larger blue flowers)

SMOOTH ASTER, *A. laevis* / Aug.–Oct. / 1'–3' / Dry to medium prairies, dry woods / Leaves blue-green with no hairs; stem leafy
(With similar blue flowers: *A. azureus*—leaves rough, of dry prairies; *A. shortii*—leaves long-triangular)

HEATH ASTER, *A. ericoides* / Aug.–Oct. / 6"–36" / Dry meadows, prairies, sandy soil / Pyramid-topped; upper leaves tiny, numerous

MARSH ASTER, *A. simplex* / Aug.–Oct. / 2'–6' / Meadows, marshes, shores / Heads numerous; in patches; leaves long, narrow; stem leafy
(*Boltonia latisquama:* heads few; no seed parachutes; plant hairless)

FROST ASTER, *A. pilosus* / Aug.–Nov. / 1'–5' / Dry prairies, disturbed ground / Clumps; arching branches; stem long-haired

BOG ASTER, *A. junciformis* / July–Oct. / 6"–32" / Bogs, wet meadows, shores / Heads few—white, pink, purple; leaves narrow, held close to stem

CALICO ASTER   hd. ⅜"

FALL ASTER   hd. ¾"

SILKY ASTER   hd. ¾"

REDSTEM ASTER   hd. 1"

NEW ENGLAND ASTER
hd. 1"

SAVORY-LEAF ASTER
hd. ¾"

DAISY FAMILY, Aster-like tribes

CALICO ASTER, A. lateriflorus /
Aug.–Oct. / 1'–4' / Dry to moist
fields, clearings, shores / In clumps;
heads on leafy horizontal or arching
branches
(CROOKED-STEM ASTER, A. prenan-
thoides: upper stem zig-zag; wet to
medium woods; rays pale blue)

REDSTEM ASTER, A. puniceus /
Aug.–Oct. / 1'–7' / Moist woods and
forest edges, wet meadows, streambanks / Often in patches; stems
leafy, red or red-striped, hairy or
smooth; leaves clasp stem; rays pale
blue to violet

**NEW ENGLAND ASTER, A. novae-
angliae** / Aug.–Oct. / 1'–7' / Medium to wet meadows, prairies /
In small patches; rays numerous,
slender—white, pink, purple; leaves
clasp stem; stem leafy

FALL ASTER, A. oblongifolius /
Sept.–Oct. / 4"–40" / Dry prairies,
cliffs / In patches; leaves small, stiff

SILKY ASTER, A. sericeus / Aug.–
Oct. / 8"–24" / Dry open woods
and prairies, cliffs / Leaves whitened
with silvery hairs; stems wiry

**SAVORY-LEAF ASTER, A. linariifo-
lius** / July–Oct. / 4"–40" / Dry
prairies, sandy woods, cliffs / In
clumps; leaves narrow, stiff, many
(ASTERS: fall bloom, leafy flower stalks,
discs turn reddish. FLEABANES: spring
bloom, discs remain yellow.)

STIFF GOLDENROD, Solidago rigida
/ July–Oct. / 1'–5' / Dry prairies
and open woods / In clumps; leaves
stiff, gray-green, rough-hairy

GOLDENROD, S. riddellii / Aug.–
Oct. / 1½'–3½' / Wet prairies and
meadows, usually limy / Leaves
sickle-shaped, without hairs
(S. ohioensis: lower leaves flat, blunt-
tipped)

**GRASSLEAF GOLDENROD, S. gram-
inifolia** / July–Oct. / 1'–4' / Meadows, prairies, roadsides, shores /
In patches; leaves very narrow, numerous on stem

STIFF GOLDENROD fl. arr. 4"

GOLDENROD fl. arr. 5"

**GRASSLEAF GOLDENROD
fl. arr. 1¼"**

SILVER ROD col. 10″

ZIG-ZAG GOLDENROD col. 1″ wide

BOG GOLDENROD col. 7″

HAIRY GOLDENROD col. 10″

SHOWY GOLDENROD col. 10″

124—125

DAISY FAMILY, Aster-like tribes

SILVER ROD, *Solidago bicolor* / Aug.–Oct. / 4″–40″ / Dry woods, sunny places, often rocky / Rays whitish

HAIRY GOLDENROD, *S. hispida* / July–Oct. / 4″–40″ / Rocky places near woods, forests, shores (*S. sciaphila:* without hairs; cliffs and sands, west)

BOG GOLDENROD, *S. uliginosa* / July–Sept. / 1′–5′ / Bogs, wet meadows and forest edges / Column often bent; leaves narrow, not hairy

SHOWY GOLDENROD, *S. speciosa* / Aug.–Oct. / 8″–60″ / Dry to medium prairies and open woods, inland sands / In clumps; leaves narrow to broad, not hairy; stem very leafy

ZIG-ZAG GOLDENROD, *S. flexicaulis* / Aug.–Sept. / 8″–48″ / Medium woods and forests / Flower arrangement at leaf-stem juncture; leaves sharp-toothed (*S. caesia:* stem round, with whitish bloom)

Goldenrods **ARCHING**

above and left
CANADA GOLDENROD
fl. arr. 10"

ELMLEAF GOLDENROD   fl. arr. 5"

MISSOURI GOLDENROD
fl. arr. 4"

DYER'S WEED
fl. arr. 14"

ELMLEAF GOLDENROD, _S. ulmifolia_
/ July–Oct. / 1'–5' / Dry woods /
Leaves coarsely toothed; heads only
at _ends_ of leafy branches

DYER'S WEED, _S. nemoralis_ / Aug.–
Oct. / 4"–40" / Dry open woods and
prairies, sand and rock soil / In
clumps; leaves narrow, gray-green;
flowers cover upper ⅓ of plant

**MISSOURI GOLDENROD, _S. mis-
souriensis_** / July–Aug. / 1'–3' / Dry
prairies and woods, sandy places /
In patches, often only a few flowering stems from the many rosettes;
upper stem not leafy or hairy
(_S. juncea:_ in clumps; leaves mostly
toward the base, but lack the 3 parallel
veins of Missouri Goldenrod)

**CANADA GOLDENROD, _S. canaden-
sis_** /July–Oct. / 1'–7' / Open woods,
meadows, prairies, shores / Photo
at left shows patch-growth and leafy
stems; plant finely-hairy
(_S. gigantea:_ not hairy; stem may have
whitish bloom; _S. patula:_ leaves oval,
upper stem 4-sided)

PEARLY EVERLASTING   hd. ⅜″

SWEET EVERLASTING   hd. ¼″

PUSSY TOES   hd. ¼″

DAISY FAMILY, Aster-like tribes

PEARLY EVERLASTING, *Anaphalis margaritacea* / July–Sept. / 1′–3′ / Forest and woods edges, fields / In patches; leaves usually white on both sides; stems very leafy

SWEET EVERLASTING, *Gnaphalium obtusifolium* / July–Oct. / 4″–32″ / Sandy prairies and disturbed ground, cliffs / Plant solitary, aromatic; leaves white beneath only
(*G. uliginosum:* heads, brown on back, almost buried in upper leaves; plant 2″–12″; north)

PUSSY TOES, *Antennaria neglecta* / April–July / 1″–4″ / Dry fields, open woods, prairies, sand / In dense patches or mats; leaves white-hairy, with only mid-vein distinct
(*A. plantaginifolia:* leaves very broadly oval, 3–7 prominent long ribs beneath; *A. parlinii:* upper leaf surfaces soon bright green, losing hairs)

INDIAN-PLANTAIN, *Cacalia suaveolens* / July–Aug. / 2′–5′ / Wet meadows, streambanks / Leaves sharply-triangular

PALE INDIAN-PLANTAIN, *C. atriplicifolia* / July–Sept. / 3′–9′ / Open woods, moist to dry prairies / Leaves white beneath, rounded-triangular
(*C. muhlenbergii:* leaves green *both* sides; moist woods and clearings)

TUBEROUS INDIAN-PLANTAIN, *C. tuberosa* / July–Aug. / 2′–5′ / Wet limy meadows, medium prairies / Only upper leaves somewhat lobed; basal leaves oval, parallel veined—often perforated by insects

BURNWEED, *Erechtites hieracifolia* / July–Oct. / 2″–80″ / Cliffs, moist clearings, meadows, and marshes / Often appears after fires; plant resembles a light green Wild Lettuce

SWEET COLTSFOOT, *Petasites frigidus* / April–June / 4″–20″ / Wet to medium forests and meadows / In patches; leaf, to 10″, palmately-lobed, whitened beneath, appears after flowers
(COLTSFOOT, *Tussilago farfara:* head solitary, yellow; summer leaves rounded or heart-shaped)

TUBEROUS INDIAN-PLANTAIN
fl. arr. 6″

INDIAN-PLANTAIN
fl. arr. 6″

**PALE
INDIAN-PLANTAIN**
fl. arr. 4½″

BURNWEED hd. ½″

SWEET COLTSFOOT
hd. ½″

left, summer leaf

PHOTOGRAPHY CREDITS

The authors are most grateful for the use of slides from: Lloyd Beesley—*Indian Cucumber Root, Nodding Wild Onion, Starry Chickweed, Larkspur, Wild Senna, Green Violet, Harbinger of Spring, Jacob's Ladder, Corn Gromwell, Horse Nettle, Justicia, Leafcup, Fleabane, Swamp Thistle, Cynthia;* Mrs. Wayne Cole— *Lion's Foot;* G. R. Cooley—*False Hellebore;* W. H. DeKarske—*Water Shield;* L. E. Fredin—*Tick Trefoil;* Dr. Robert French—*Water Smartweed, Woolly Blue Violet;* Miss Joan Gibson—*Blue-eyed Grass;* H. H. Hadow —*Water-Plantain, Dewdrops, Seneca Snakeroot, Forget Me Not, Fog Fruit;* Dr. T. G. Hartley—*Gold Hedge-hyssop;* Bernard Horne—*Yucca, Yellow Twayblade, Three Birds, Lizard Tail, Corn Cockle, Yellow Wild Indigo, Hog Peanut, Dotted St. John's-wort, Fireweed, Gaura, Swamp Candles, Columbo, Moth Mullein, Squaw Root, Beech Drops, Buckhorn, Silver Rod, Elecampane, Beach Thistle;* Miss Harriet A. Irwin—*Showy Ladyslipper, Pink Ladyslipper;* Mrs. B. E. Kline—*Beach Pea;* K. I. Lange—*Club-spur Orchid, Yellow Fringed Orchid, Grass Pink, Meadow Beauty, Seed Box, Pipsissewa, Spotted Pipsissewa, Waterleaf, Pink Gerardia, Monkshood;* R. E. Lee—*Prairie Violet, Longleaf Bluets;* R. J. Lukes—*Cancer-root, Naked Miterwort;* Dr. Elizabeth Lunn—*Wild Indigo;* L. C. McDowell—*Devil's Bit, Whorled Pogonia, 3-leaf Sedum, Bush Clover, Early Yellow Violet, Morning Glory, Hound's Tongue, Horse Balm, Bluets;* from the estate of Angus McVicar— *Kitten Tail;* A. M. Peterson—*Sullivantia;* Dr. J. S. Pringle—*Yellow Flag, Poison Ivy leaves;* Edward Prinz— *Starry Campion, Grass of Parnassus, Prairie Loosestrife, Bottle Gentian, Black-eyed Susan;* Miss Joan Rahn—*Fire Pink, Poison Ivy flowers, Cocklebur;* J. H. Reeder—*Butterwort;* Sam Ruegger—*Fringed Gentian, Sweet Coltsfoot flower, Bogbean, Indian Paint Brush, Partridge Berry;* Valdemar Schwarz—*Hooker's Orchid, Dewberry, Partridge Pea, Rattle Box, Bitterbloom, Elmleaf Goldenrod;* Mr. and Mrs. C. M. Smith—*Heartleaf Twayblade, Pinesap, Clammy Ground Cherry, Helleborine, Crinkleroot;* Dr. E. G. Voss—

Pipewort, Flowering Rush, Adder's Mouth, Water Willow, Sand Cancer-root; Mrs. H. M. Williams—*Sundew, Gromwell, Hoary Vervain, Indian Physic, Goat's Beard;* J. F. Zoerb—*Lead Plant, Poppy Mallow.*

All other color photographs are by the authors.
Drawings by Booth Courtenay.

INDEX TO CHARTS AND COLOR SECTION